Praise for

CAESAR'S LAST BREATH

Selected as one of the Best Science Books of the Year by
Science News and *The Guardian*
Finalist for the AAAS/Subaru SB&F Prize

"Informative and well organized...Kean's ability to explain with clear, vivid analogies provides diverse readers access to previously remote scientific concepts." — Andrea Jenney, *Science*

"*Caesar's Last Breath* brims with such fascinating tales of chemical history that it'll change the very way you think about breathing...Kean crams the book full of wild yarns told with humorously dramatic flair...The effect is oddly intimate, the way all good storytelling is—you feel like you're sharing moments of geeky amusement with a particularly hip chemistry teacher." — Chelsea Leu, *San Francisco Chronicle*

"With sly wit and boyish wonder, Kean's vignettes about key events in the understanding of air include numerous entertaining detours, such as the work of 'William McGonagall, probably the worst poet who ever lived,' and Le Pétomane, a flatulence artist." — Gemma Tarlach, *Discover*

"Compelling stuff, written with verve and in a style that veers between simple lightheartedness and open jocularity...Eminently accessible and enjoyable. A real gas, in short." — Robin McKie, *The Observer*

"A page-turner...Kean clearly delights in weaving in the unexpected...a compelling narrative of geologic history."
— Erika Engelhaupt, *Science News*

"Kean illuminates the science in everything from Earth's vaporous origins to the function of hydrogen in early aeronautic balloons and the ammonia and butane that Albert Einstein and Leo Szilard harnessed in the 1930s for a novel refrigerator."
— Barbara Kiser, *Nature*

"Entertaining...Kean packs *Caesar's Last Breath* with personalities and human interest...There is no denying the pleasure and indeed the wealth of scientific information to be obtained from reading *Caesar's Last Breath*." — Clive Cookson, *Financial Times*

"Riveting...Kean has a knack for distilling chemistry to its essential elements, using stories and humor...For anyone interested in the invisible forces that sustain life, this is a dose of fresh air."
— Chad Comello, *Library Journal*

"Kean pumps chemical and historical trivia into this tale about air and the gases of which it is composed...Entertaining."
— *Publishers Weekly*

"Richly informative...Once again, Kean proves his mettle as one of science literature's most gifted practitioners."
— Carl Hays, *Booklist*

"In *Caesar's Last Breath,* Kean takes on the science of gases in his trademark effervescent, loopy style...A lively, rewarding journey through the evolution of Earth's gaseous atmosphere."
— Bruce Jacobs, *Shelf Awareness*

"With fizzy humor and an exuberant enthusiasm for scientific ephemera, Kean entices us to explore the alchemy of air and atmosphere...*Caesar's Last Breath* is a rollicking, zigzag romp through the science of air—one that gives us pause to consider our immortality beyond an Earthly existence: 'Some tiny bit of you—molecules that danced inside your body...could live on in a distant world.'"
                              —Alison Hood, *BookPage*

"An engaging summer read and worth becoming entangled."
                    —Sheilla Jones, *Winnipeg Free Press*

"Sam Kean has done it again—this time clearly and entertainingly explaining the science of the air around us. He is a gifted storyteller with a knack for finding the magic hidden in the everyday."
                         —Daniel H. Pink, author of *Drive*

"The most fun to be had from nonfiction is a good science book, with a writer of craft who can capture both the excitement and the elegance of science, the incredible fact that this is really how it works. Sam Kean is such a writer, and *Caesar's Last Breath* is such a book. An enormous pleasure to read."
                   —Mark Kurlansky, author of *Paper* and *Salt*

"Fascinating stories, so insightful, informative, and disarmingly written. It gave this astronaut a new respect for the air around us all, and made me delightfully more aware of each breath I take."
                              —Col. Chris Hadfield, author of
                      An Astronaut's Guide to Life on Earth

"It's a vast and misty topic, but this vibrant book makes the chemistry of air riveting."
                      —James McConnachie, *The Times* (UK)

Also by Sam Kean

The Tale of the Dueling Neurosurgeons

The Violinist's Thumb

The Disappearing Spoon

CAESAR'S LAST BREATH

BREATH

And Other True Tales of History, Science, and the
Sextillions of Molecules in the Air Around Us

——◆——

SAM KEAN

BACK BAY BOOKS
LITTLE, BROWN AND COMPANY
New York   Boston   London

Back Bay Books / Little, Brown and Company
Hachette Book Group
1290 Avenue of the Americas, New York, NY 10104
littlebrown.com

Originally published in hardcover by Little, Brown and Company, July 2017
First Back Bay paperback edition, June 2018

Back Bay Books is an imprint of Little, Brown and Company, a division of Hachette Book Group, Inc. The Back Bay Books name and logo are trademarks of Hachette Book Group, Inc.

The publisher is not responsible for websites (or their content) that are not owned by the publisher.

The Hachette Speakers Bureau provides a wide range of authors for speaking events. To find out more, go to hachettespeakersbureau.com or call (866) 376-6591.

ISBN 978-0-316-38164-2 (hc) / 978-0-316-38165-9 (pb)
Library of Congress Control Number: 2016954781

10 9 8 7 6 5 4 3 2

LSC-C

Printed in the United States of America

To see a World in a Grain of Sand
And a Heaven in a Wild Flower,
Hold Infinity in the palm of your hand
And Eternity in an hour.

— WILLIAM BLAKE,
"AUGURIES OF INNOCENCE"

Contents

Introduction: The Last Breath   3

I. MAKING AIR: OUR FIRST FOUR ATMOSPHERES

Chapter One: Earth's Early Air   17

Interlude: The Exploding Lake   43

Chapter Two: The Devil in the Air   49

Interlude: Welding a Dangerous Weapon   75

Chapter Three: The Curse and Blessing of Oxygen   81

Interlude: Hotter than the Dickens   111

II. HARNESSING AIR: THE HUMAN RELATIONSHIP WITH AIR

Chapter Four: The Wonder-Working Gas of Delight   121

Interlude: Le Pétomane   151

Chapter Five: Controlled Chaos   158

Interlude: Steeling Yourself for Tragedy   185

Chapter Six: Into the Blue   196

Interlude: Night Lights   223

III. FRONTIERS: THE NEW HEAVENS

Chapter Seven: The Fallout of Fallout   231 x • Contents

Interlude: Albert Einstein and the People's Fridge   259

Chapter Eight: Weather Wars   268

Interlude: Rumbles from Roswell   293

Chapter Nine: Putting on Alien Airs   304

Acknowledgments   327

Notes and Miscellanea   329

Works Cited   357

Index   363

CAESAR'S
LAST BREATH

The Last Breath

Indulge me in a modest experiment. For the next few seconds try paying close attention to the air escaping your body, as if this were your last living breath on Earth. How much do you really know about this air? Feel your lungs deflate and sag inside your chest. What's really going on inside there? Put your hand in front of your lips and feel how the gas escaping them has transformed inside you, growing warmer and more humid, perhaps acquiring an odor. What sort of alchemy caused that? And although your sense of touch isn't nearly discriminating enough, imagine that you can feel the individual molecules of gas pinging your fingertips, impossibly tiny dumbbells caroming off into the air around you. How many are there, and where do these molecules go?

Some don't get far. As soon as you take another breath, they come rushing back into your lungs, like waves that fling themselves onto the shore before being retracted by the sea. Others stray a little farther and make a break for freedom in the next room before returning as well, miniature prodigal sons. Most simply join the anonymous masses of the atmosphere and begin to spread around the globe. But even then, perhaps months later, a few weary pilgrims will stagger back to you. You might be a very different person between your first and second encounters with these molecules, but the ghosts of breaths past continue to flit around you every second of every hour, confronting you with every single yesterday.

Of course, you're not alone in experiencing this; the same thing happens to every other person on Earth. Moreover, your ghosts are almost certainly entangled with theirs, since they almost certainly inhaled and expelled and rebreathed a few of those very same molecules after you did—or even before you did. In fact, if you're reading this in public, you're inhaling the exhaust from everyone around you right now—secondhand breath. Your reaction to this will probably depend on the company you keep. Sometimes we enjoy this mingling of airs, as when lovers lean in and we feel their breath on our necks; sometimes we abhor it, as when the chatterbox next to us on the plane has eaten garlic for lunch. But short of breathing from a tank, we can't escape the air of those around us. We recycle our neighbors' breaths all the time, even distant neighbors'. Just as light from distant stars can sparkle our irises, the remnants of a stranger's breath from Timbuktu might come wafting in on the next breeze.

Even more startling, our breaths entangle us with the historical past. Some of the molecules in your next breath might well be emissaries from 9/11 or the fall of the Berlin Wall, witnesses to World War I or the star-spangled banner over Fort McHenry. And if we extend our imagination far enough in space and time, we can conjure up some fascinating scenarios. For instance, is it possible that your next breath—this one, right here—might include some of the same air that Julius Caesar exhaled when he died?

You know the story. The ides of March, Rome, 44 BC. Julius Caesar—*pontifex maximus, dictator perpetuo,* the namesake of July and the first living Roman to have his picture on a coin—enters the Senate meeting hall, looking surprisingly spry after a rough night. At a dinner party he'd attended, conversation had strayed into the rather morbid topic of the best way to die. (Caesar had declared his preference for a sudden, unexpected end.) An epileptic, he'd also slept poorly that night, and his wife had suffered ominous dreams about their house collapsing and her holding a bloodied Caesar in her arms.

As a result of all this, he almost stayed home that morning. But at the last minute he ordered his servants to ready his litter, and as his retinue made their way toward the Forum, he finally relaxed, his breath coming freer and easier. He even teased a soothsayer along the way, a man who, a month before, had prophesied doom for Caesar sometime before mid-March. Caesar filled his lungs and shouted, "The ides of March have come!" The seer answered without smiling. "Aye, Caesar, but not passed."

As Caesar entered the meeting hall, hundreds of senators rose to their feet. It was likely stuffy in there, as their mingled breath and body heat had been warming the air for some time. Before Caesar could settle into his golden chair, though, a senator named Cimber approached him with a petition asking for pardon for Cimber's brother. Cimber knew that Caesar would never grant this, but that was the point. Cimber kept begging and Caesar kept refusing, and sixty other senators now had a chance to creep forward, as if offering support. Caesar sat in the midst of them, imperial and increasingly irritated. He tried to cut off discussion, but Cimber cupped his hands on Caesar's shoulders as if to plead with him—then yanked his purple toga down, exposing Caesar's breast.

"Why, this is violence," Caesar said. He had no idea how right he was. A senator named Casca lunged with his dagger a moment later, gashing Caesar's neck. "Casca, you villain, what are you doing?" Caesar cried, still more confused than angry. But as the crowd of "petitioners" pressed in, each man pulled aside his toga, exposing a bit of skin, and opened the leather pouch on his belt where he

The Death of Caesar, by Vincenzo Camuccini.

normally kept a stylus. Rather than sixty pens, sixty iron daggers emerged. Caesar finally understood. *Sic semper tyrannis.*

Caesar fought back at first, but after the first few stabs the marble floor beneath his sandals grew slippery with blood. He soon got tangled in his garment and fell. At this the assassins pounced, stabbing Caesar twenty-three times in all. In looking over the body later, Caesar's doctor determined that twenty-two of the gashes were superficial. To be sure, his body would have panicked a little more with each wound, and the shock would have withdrawn blood from periphery to core, to keep oxygen flowing to his vital organs. But he still would have survived, the doctor said, if not for one of the cuts: a single stab to the heart.

According to most accounts, Caesar wrapped himself in his toga before falling, and died without a whimper. But according to one account—and it's easy to see why this account above all has captivated people for two thousand years—Caesar felt a stab in his groin just before going down, and wiped his blood-smeared eyes. In doing so he spotted his protégé Brutus amid the pack, his dagger gleaming red. Caesar took this in and murmured, "You, too, my son?," half question, half answer. He then covered himself to preserve a little dignity, and crumpled to the floor with a final, pained gasp.

So what "happened" to that breath? At first the answer seems obvious: it's gone. Caesar died so long ago that little remains of the building where he fell, much less of his body, which was cremated into ash. Even the iron daggers have likely disintegrated by now, rusting into scabs of dust. So how could something as ephemeral as a breath still linger? If nothing else, the atmosphere extends so far and wide that Caesar's last gasp has surely been dissolved into nothingness by now, effaced into the æther. You can open a vein into the ocean, but you don't expect a pint of blood to wash ashore two thousand years later.

I mean, consider the numbers. Your lungs expel a half liter of air with every normal breath; a gasping Caesar probably exhaled a full liter, a volume equivalent to a balloon five inches wide. Now compare that balloon to the sheer size of the atmosphere. Depending on where you cut it off, the bulk of the atmosphere forms a shell around Earth about ten miles high. Given those dimensions, that shell has a volume of two billion cubic miles. Compared to the atmosphere at large, then, a one-liter breath represents just 0.00000000000000000001 percent of all the air on Earth. Talk about tiny: Imagine gathering together all of the hundred billion people who ever lived—you, me, every last Roman emperor and pope and Dr. Who. If we let those billions of people stand for the atmosphere, and reduce our population by that percentage, you'd have just 0.00000000001 "people" left, a speck of a few hundred cells, a last breath indeed. Compared to the atmosphere, Caesar's gasp seems like a rounding error, a cipher, and the odds of encountering any of it in your next breath seem nil.

Before we shut the door on the possibility, though, consider how quickly gases spread around the planet. Within about two weeks, prevailing winds would have smeared Caesar's last breath all around the world, in a band at roughly the same latitude as Rome — through the Caspian Sea, through southern Mongolia, through Chicago and Cape Cod. Within about two months, the breath would cover the entire Northern Hemisphere. And within a year or two, the entire globe. (The same holds true today, naturally — any breath or belch or exhaust fume anywhere on Earth will take roughly two weeks, two months, or one or two years to reach you, depending on your relative location.)

Surely, though, wouldn't those winds have spread the breath so thin that nothing remained? Wouldn't the breadth of the breath erase it? Perhaps not. In the discussion above, we treated Caesar's breath as a single mass, a single thing. But if we drill down to the itty-bitty, this singular mass of air pixelates out into discrete molecules. So while on some level (the human level) Caesar's last breath does seem to have disappeared into the atmosphere, on a microscopic level his breath hasn't disappeared at all, since the individual molecules that make it up still exist. (Despite how "soft" air seems, most air molecules are pretty hardy: the bonds that bind their atoms together are some of the strongest in nature.) So in asking whether you just inhaled some of Caesar's last breath, I'm really asking whether you inhaled any *molecules* he happened to expel at that moment.

The answer of course depends on how many molecules we're talking. And with a bit of freshman chemistry, you can calculate that one liter of air at any sort of reasonable temperature and pressure corresponds to approximately 25 sextillion (25,000,000,000,000, 000,000,000) molecules. That's a stupefying number, big beyond comprehension. Imagine Bill Gates cashing out his entire $80 billion fortune, converting it to $1 bills, and then stuffing them all under his mattress. Imagine then that he withdraws every individual dollar, one by one, and uses each one as seed money to found another software company. Now imagine that each of those 80 billion companies goes off like gangbusters and yields a return of $80 billion on its own. Add all that cash together—80 billion times 80 billion—and you're still four times short of the number of molecules you inhale with every breath. All the world's roads and all the world's canals and all the world's airports in the history of humankind haven't handled nearly as much traffic as our lungs do every second. From this perspective Caesar's last breath seems innumerable, and it seems inevitable that you'd inhale at least a few molecules of it in your next breath.

So which number wins out? The gargantuan number of molecules that Caesar expelled, or the insignificance of any single breath compared with the atmosphere? To answer that, it might help to consider an analogous situation, a prison break and a manhunt.

Say that all 300 inmates of Alcatraz at its peak—Al Capone, Robert "Birdman" Stroud, George "Machine Gun" Kelly, and 297 close friends—overwhelmed the guards, snagged some boats, and escaped to the mainland. Say also that, being street-smart, the fugitives flee San Francisco and (much like a gas) diffuse across the entire United States to lower their chances of capture. Say finally that you're a bit paranoid about all this, and you want to know whether any fugitive will likely wander into your hometown. Are your fears justified?

Well, the United States covers 3.8 million square miles. Given 300 inmates, that works out to around one fugitive per 125,000 square miles. My hometown in South Dakota sprawls across roughly 75 square miles of prairie, so the number of Alcatraz escapees we could expect there—divide 75 by 125,000—is 0.006. In other words, zero. We can't be sure it's zero, because one might randomly show up. But in all likelihood Alcatraz simply couldn't have flooded the country with enough thugs to make my hometown a probable sanctuary.

There are bigger prisons than Alcatraz, though. Imagine the same scenario unfolding with Cook County Jail in Chicago, which holds 10,000 inmates. Because more prisoners would be flooding the country, the odds of one matriculating in my hometown would rise to around 20 percent. Still not a certainty, but suddenly I'm sweating. The odds would of course rise still higher if the entire U.S. prison population (an incredible 2.2 million people) all escaped at once. This time the number of convicts on the lam in my hometown would jump to 43 — not percent, but 43 actual fugitives. With Alcatraz, in other words, the tininess of my hometown within the vast United States kept it safe. But in an apocalyptic, nationwide prison break, the sheer number of escapees would overwhelm that tininess and all but ensure that some of the outlaws would take refuge there.

With that in mind, consider Caesar's final breath again. The molecules of air escaping his lungs are the prisoners escaping their cells. Their spread across the country is the diffusion of gas molecules into the atmosphere. And the odds of a prisoner ending up in a (relatively tiny) given town are the odds of any one molecule getting swept up in your (relatively tiny) next breath. So the question becomes: Is Caesar's final breath like Alcatraz, spilling too few molecules into the air to make a difference? Or is it like the entire U.S. prison population escaping, making it a statistical certainty?

Somewhere in between. Sort of like matter meeting antimatter, the 25,000,000,000,000,000,000,000 molecules and the 0.00000000000000000001 percent almost exactly cancel each other out. When you crunch the numbers, you'll find that roughly one particle of "Caesar air" will appear in your next breath. That number might drop a little depending on what assumptions you make, but it's highly likely that you just inhaled some of the very atoms Caesar used to sound his cri de coeur contra Brutus. And it's a certainty that, over the course of a day, you inhale thousands.

Think about that. Across all that distance of time and space, a few of the molecules that danced inside his lungs are dancing inside yours right now. And given how often we breathe (once every four seconds), this happens 20,000 times every day. Over the years you might even incorporate some of them into your body. Nothing liquid or solid of Julius Caesar remains. But you and Julius are practically kissing cousins. To misquote a poet, the atoms belonging to his breath as good as belong to you.

———<o>———

Mind you, there's nothing special about Caesar, either. I've heard variants of the "Caesar's breath" problem that used Jesus on the cross as the protagonist (I went to Catholic school), and really, you could pick anyone who suffered through an agonizing last breath: the masses at Pompeii, Jack the Ripper's victims, soldiers who died during gas attacks in World War I. Or I could have picked anyone who died in bed, whose last breath was serene—the physics is identical. Heck, I could have picked Rin Tin Tin or Jumbo the giant circus elephant. Think of anything that ever breathed, from bacteria to blue whales, and some of his, her, or its last breath is either circulating inside you now or will be shortly.

Nor should we limit ourselves to stories about breathing. The how-many-molecules-in-X's-last-breath exercise has become a classic thought experiment in physics and chemistry courses. But whenever I heard someone rattle on about so-and-so's last breath, I always got restless. Why not be more audacious? Why not go further and trace these air molecules to even bigger and wilder phenomena? Why not tell the full story of *all* the gases we inhale?

Every milestone in Earth's history, you see—from the first Hadean volcanic eruptions to the emergence of complex life—depended critically on the behavior and evolution of gases. Gases not only gave us our air, they reshaped our solid continents and transfigured our liquid oceans. The story of Earth *is* the story of its gases. Much the same can be said of human beings, especially in the past few centuries. When we finally learned to harness the raw physical power of gases, we could suddenly build steam engines and blast through billion-year-old mountains in seconds with explosives. Similarly, when we learned to exploit the chemistry of gases, we could finally make steel for skyscrapers and abolish pain in surgery and grow enough food to feed the world. Like Caesar's last breath, that history surrounds you every second: every time the wind comes clattering through the trees, or a hot-air balloon soars overhead, or an unaccountable smell of lavender or peppermint or even flatulence wrinkles your nose, you're awash in it. Put your hand in front of your mouth again and feel it: we can capture the world in a single breath.

That's the goal of *Caesar's Last Breath* — to make these invisible stories of gases visible, so you can see them as clearly as you can see your breath on a crisp November morning. At various points in the book we'll swim with radioactive pigs in the ocean and hunt insects the size of dachshunds. We'll watch Albert Einstein struggle to invent a better refrigerator, and we'll ride shotgun with pilots unleashing top-secret "weather warfare" on Vietnam. We'll march with angry mobs, and be buried inside an avalanche of vapors so hot that people's brains boiled inside their skulls. All of these tales pivot on the surprising behavior of gases, gases from lava pits and the guts of microbes, from test tubes and car engines, from every corner of the periodic table. We still breathe most of them today, and each chapter in this book picks one of them as a lens to examine the sometimes tragic, sometimes farcical role that gases played in the human saga.

The book's first section, "Making Air: Our First Four Atmospheres," covers gases in nature. This includes the formation of our very planet from a cloud of space gas 4.5 billion years ago. Later a proper atmosphere emerged on our planet, as volcanoes began expelling gases from deep inside Earth. The emergence of life then scrambled and remixed this original atmosphere, leading to the so-called oxygen catastrophe (which actually worked out pretty well for us animals). Overall the first section explains where air comes from and how gases behave in different situations.

The second section, "Harnessing Air: The Human Relationship with Air," examines how human beings have, well, harnessed the special talents of different gases over the past few centuries. We normally don't think of air as having much mass or weight, but it does: if you drew an imaginary cylinder around the Eiffel Tower, the air inside it would weigh more than all the metal. And because air and other gases have weight, they can lift and push and even kill. Gases powered the Industrial Revolution and fulfilled humanity's ancient dream of flying.

The book's third section—"Frontiers: The New Heavens"—explores how our relationship with air has evolved in the past few decades. For one thing, we've changed the composition of what we breathe: the air you inhale now is not the same air your grandparents inhaled in their youth, and it's markedly different from the air people breathed three hundred years ago. We've also started to explore the atmospheres of planets beyond our solar system, opening up the possibility that our descendants could leave Earth altogether and start over on a planet filled with gases we can't even imagine yet.

In addition to these big stories, the book also contains a series of vignettes, collectively called "Interludes." They expand on the themes and ideas in the main chapters and explain the role that gases play in phenomena like refrigeration, home lighting, and intestinal distress. (Just for kicks, a few vignettes also stray into some not-so-everyday topics, such as spontaneous combustion and the Roswell alien "invasion.") Many of the gases featured in these sections are trace components of air—compounds that make up just a few parts per million, or a few parts per billion, of what we breathe. But in this context *trace* doesn't mean insignificant. Think of a glass of wine. Wine is well over 99 percent water and alcohol, but water and alcohol alone do not a wine make. Wines have scores of other flavors—hints of leather, chocolate, musk, plums, and so on. Just so, trace gases in the air add overtones and finish to the air we breathe and the stories we can tell.

—◄○►—

If you ask people on the street what air is, you often get wildly different accounts, depending on which gases they focus on or whether they're talking about air on a microscopic or macroscopic level. That's fine: air is big enough to accommodate all those points of view. In fact, I hope this book compels you to revise your own mental picture of air, and I think your notion of air will indeed shift from chapter to chapter, leaving you with a more holistic view of it.

It's worth asking yourself what you think of air, too, because air is the single most important thing in your environment right now. You can survive without food, without solids, for weeks. You can survive without water, without liquids, for days. Without air, without gases, you'd last a few minutes at most. I'll wager, though, that you spend the least amount of time thinking about what you're breathing. *Caesar's Last Breath* aims to change that. Pure air is colorless and (ideally) odorless, and by itself it sounds like nothing. That doesn't mean it's mute, that it has no voice. It's burning to tell its story. Here it is.

I. Making Air

———⟨o⟩———

OUR FIRST
FOUR ATMOSPHERES

In "Making Air," we'll examine two major questions about air: where our atmosphere came from, and what its major ingredients are. Overall, Earth has had several distinct atmospheres in its history, each with a unique mix of gases. Many of those gases ultimately came from volcanoes, and some date back to the very early days of our planet, long before life existed. But life has remade and reworked the atmosphere in several ways since then, especially by adding oxygen.

CHAPTER ONE

Earth's Early Air

Sulfur dioxide (SO_2)—currently 0.00001 parts per million in the air; you inhale 120 billion molecules every time you breathe

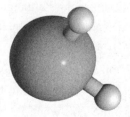

Hydrogen sulfide (H_2S)—currently 0.000005 parts per million; you inhale 60 billion molecules every time you breathe

Fear of being murdered first drove Harry Truman into hiding in the foothills of Mount Saint Helens in 1926. Not *that* Harry Truman, although this one—Harry Randall Truman—did appreciate his namesake. "He's a gutsy old codger," said Truman of Truman. "Bet he'll go down as one of the greatest goddamn presidents." This Truman would know a thing or two about being a gutsy old codger. After fleeing to Washington State at age thirty, he stuck out fifty-four years of brutal, isolating winters under the glare of Mount Saint Helens. And even when the mountain began to steam and snort and bellow in the spring of 1980, it couldn't dislodge him except in the most spectacular way possible—by blowing him straight up into the atmosphere.

Truman's family of loggers had moved to Washington State during his childhood, and after he graduated from high school he enlisted in the army, serving as an airplane mechanic during World War I. (A born raconteur, Truman would later claim that he flew combat missions overseas, his white scarf streaming behind him in the open cockpits of the day.) Back home he married a sawmill owner's daughter and became a car mechanic, but he found both marriage and regular employment tedious. He tried prospecting for gold instead and found it worse than tedious, an outright pain in the ass.

So when Prohibition descended, he started bootlegging, a job more suited to his temperament. Playing fast and loose with the law tickled him, and he liked the quick cash. He also enjoyed a drink now and then, and didn't appreciate a gaggle of do-gooders lecturing him on the evils of whiskey. Eventually he partnered with some gangsters in Northern California and started running hooch up the coast, supplying whorehouses and speakeasies along the way. He was having a great goddamn time of it all, but in 1926, something spooked him. He never quite said what. Perhaps he got too friendly with someone's special lady or tried to horn in on some mobster's territory. Regardless, he started carrying a submachine gun around.

One day he finally grabbed his wife and daughter and fled to the forests around Mount Saint Helens to lie low.

Motormouth Harry Randall Truman, drinking a glass of "panther pee," at his beloved lodge in the shadow of Mount St. Helens. (Photo courtesy U.S. Forest Service)

To make do he began managing a gas station and grocery store three miles north of the summit; he gradually expanded that into a campground with cabins and boats to rent. It proved a popular location. Gorgeous fir trees, some 250 feet tall and eight feet in diameter, ringed his house. The campgrounds also contained Spirit Lake, a 2.5-mile-long slip of water as cold and clear as chilled gin. Given the remoteness, Truman could keep on bootlegging as well, and he stashed several barrels of homemade whiskey—which he labeled "Panther Pee"—around the forest.

His wife, meanwhile, found the isolation grating. Nor did she appreciate being separated from their daughter, who attended boarding school several miles distant. Perhaps inevitably, Truman and his wife divorced in the early 1930s. Truman quickly remarried in 1935, but the second Mrs. Truman—every bit as pissy and vinegary as Truman himself—didn't last much longer. (It didn't help that Truman tried to "win" arguments with her by tossing her into Spirit Lake. She couldn't swim.) So Harry tried again, first wooing another local girl and then switching allegiance to her sister, Edna. Not exactly a romantic start, but once he fell in love with Eddie, Truman never quite got the barb out of his heart.

Eddie must have loved him something fierce, too, because Truman didn't sound like the easiest husband. Most dawns, while he attended to chores, she had to rise early to whip up his favorite breakfast: scrambled cow brains with a glass of buttermilk to wash them down. "This will keep a man vi-i-irile," he'd cackle. Eddie also had to put up with his mouth. Because he spoke quickly, people sometimes had trouble understanding Truman, but you could always pick out the cuss words. Friends were "old bastards," while people he knew less well were "stupid sons of bitches." (He once drove Supreme Court justice William Douglas off his property for looking like a sissy, although they later became friends.) Truman could be a know-it-all, too. Whenever a photographer visited to snap a few pictures of Mount Saint H, Truman would butt in. "You gotta put a human being in the goddamn thing," he'd complain. "A little human interest is what means so much to the goddamn public." Listening to interviews decades later, after Truman became famous, you'd think his favorite word was *beep.*

After a good day of renting out boats on Spirit Lake, Truman might take in $1,500 in cash. He'd find Eddie in the café they ran together and hand over the money, then pour himself a tall glass of his favorite panther pee—Schenley whiskey cut with Coke—and shoot the shit with his guests. As a couple they sometimes splurged on things like Harry's pink 1956 Cadillac, which wasn't exactly practical on the mountain roads but which he loved almost half as much as he loved Eddie. More often, though, they plowed the money back into the campgrounds or café. Or they saved it for the lean months, those winters when nine feet of snow fell and money got, as guess who said, "tighter than a bull's ass at fly time." But even then, Truman couldn't help but marvel at his luck, living where he did. "Look at that!" he'd say, pointing at Mount Saint Helens. "You'll never see anything more beautiful in the whole goddamn world than that old mountain."

However correct the sentiment—Mount Saint Helens was indeed beautiful, the platonic ideal of snow-capped mountainness—this was actually bad geology on Truman's part, since the cone of Mount Saint Helens wasn't old at all; it barely existed in Julius Caesar's day. It might seem impossible for thousands of feet of mountain to rise in just a few thousand years, but Mount Saint Helens was an active volcano, and active volcanoes can add height quickly by packing lava onto their slopes. In fact, although Fire Mountain (as local Indians called it) hadn't erupted since 1857, even an amateur could spot evidence of past eruptions all around, such as hiking trails peppered with old ash, and volcanic pumice stones so porous that, when tossed into Spirit Lake, they floated. Every so often visitors would feel a tremor, too. But for Truman, the danger only added to the mountain's allure.

Truman had been living in the shadow of Mount Saint Helens for nearly fifty years when everything unraveled. One night in September 1975, Eddie felt ill and retired to bed early. Truman had friends visiting from town that night, and he telephoned them to bring up a bottle of Schenley for himself and a surprise gift for Eddie, a plant, to cheer her up. When the plant arrived, he brought it straight up to their room—and came racing back down the stairs moments later. He was talking even faster than normal, verging on incoherence. "Eddie's sick! Eddie's sick!" was all they could make out. In fact, Eddie had already died of a heart attack. But for another hour he kept begging people to help her.

Eddie's death opened up a sinkhole inside Truman. Smiles came only grudgingly afterward, and friends who'd always laughed at how spry and fit he seemed — "tougher than boiled owl," one called him — now remarked on how run-down he looked, showing every one of his seventy-nine years. During the summers he at least had the campground to distract himself. But people worried about him during those six-month-long winters, when for weeks at a stretch he might have no other companion than Mount Saint Helens — two lonely, solitary figures staring at each other across miles of forest.

That isolation was about to end, however. Around the time Eddie died, a few government geologists who had been sampling layers of rock around Mount Saint Helens announced that they'd discovered a grim record of eruptions in its past. A report they released in 1978 went even further, describing the mountain as "more active and more explosive during the last 4,500 years than any other volcano in the contiguous United States." Prophetic words. In nature, beautiful often means deadly, and that beautiful mountain cone might as well have been a cannon barrel.

Mount Saint Helens was the greatest geology lesson in American history. It also provided, believe it or not, a fascinating peek at the early days of our planet and the creation of our atmosphere. We don't usually think about air as being created — it seems like it just *is* — but all planets have to manufacture their atmospheres from scratch. And despite how nasty volcanic fumes might seem, they supplied the basic ingredients on Earth. Understanding our air, then, requires understanding these explosions of lava and gas — and there's no better place to start than the most scrutinized eruption in history and the unlikely hero who helped make it famous.

◄○►

You can trace the danger lurking beneath Mount Saint Helens to the very early days of planet Earth. Around 4.5 billion years ago, a supernova detonated in our neck of the cosmos and sent a shock wave into space. This shock wave plowed into a sea of mostly hydrogen gas that happened to be nearby and stirred something inside it to life, causing the sea to grow choppy and swirl in a vortex about its center. Gravity eventually sucked together 99.9 percent of that gas to form a new star, our sun. The majority of the remaining gas got pushed to the edges of this incipient solar system, where it formed gas giant planets like Jupiter and Saturn.

Meanwhile, a small amount of gas got stranded in between the sun and the gas giants, and the elements within this cloud—oxygen, carbon, silicon, iron, and others—began clumping together on their own, first into microscopic specks, then into boulders and asteroids, then continent-sized rocks. Gravity being an efficient janitor, this sweeping together of small bits into large masses happened quickly—a million years by some accounts. When enough pieces had bonded, several rocky planets emerged, including Earth. Everything around you, then, no matter how solid it seems—the floor beneath your feet, the book you're holding, even your body—all started off as a gas. You *are* an ex-gas.

Although it might sound dubious to talk about a solid body like Earth as an erstwhile gas, it makes sense if you examine what solids and gases fundamentally are. The molecules in a solid have a fixed address and cannot shift around much; that's why solids retain their shape so well. In liquids, the molecules still touch and rub together, but they have more energy and more freedom to slide about, which explains why liquids flow and take the shape of their containers. Gas molecules don't rub elbows at all; they're feral, with much more space between neighbors. And when gas molecules do meet, they crash together and rebound in new directions, like a chaotic game of 3-D billiards. An average molecule of air at 72°F zips around at a thousand miles per hour.

On some level, then, solids and liquids and gases seem entirely different—and early scientists in fact classified things like water, ice, and steam as distinct substances. Today we know that's not true. Heat up solid ice, and its molecules snap their chains and start to slide around like a liquid. Pump still more heat energy in, and the liquid molecules leap into the air and become a gas. It's all the same substance, just in a different guise. And other materials can undergo this transformation as well. We're not used to thinking about, say, the iron or silicon or uranium inside rocks as potential vapors, but every last substance on the periodic table can become a gas if hot enough. The reverse process works as well. Withdraw heat from a gas, and it will dew into a liquid. Subtract more heat, and a solid forms. Increasing the pressure on a gas can also crush its molecules into less sprightly states of matter.

These different states of matter mingled uneasily on the early Earth—a seething, molten mass that looked nothing like the manicured planet we know today. After (solid) space rocks began clumping together to form a planet initially, the immense gravitational pressures involved melted most of them into liquids. Denser liquids (such as molten iron) sank into the core, while lighter ones rose—only to resolidify at the surface when they encountered the cold sting of outer space. Overall this Earth resembled an egg, with a mostly iron yolk, a gooey mantle surrounding it, and a thin black shell of rock. The big difference was that this shell was fractured into millions of pieces, with molten lava leaking out between the cracks. Darkness never quite descended at night on the early Earth because of this lava's orange glow, and not infrequently a spume of it would leap into the air, like the garden fountains in hell.

This early Earth did have an atmosphere (atmosphere number one, for those keeping track), but it was so wispy it hardly deserved the name. It consisted mostly of hydrogen and helium that had been stranded between the sun and Jupiter. And not long after it formed, this atmosphere disappeared on us, as the solar wind—a nor'easter of protons and electrons that originate inside the sun—swept these gases away from Earth and into space. Many of these gas atoms escaped of their own accord as well. One law of gases says that small molecules move at much higher speeds than large, stodgy molecules. Hydrogen and helium sit atop the periodic table as the lightest and therefore swiftest elements, and every day a certain fraction of them would exceed the escape velocity of Earth (seven miles per second). Like trillions of tiny Saturn rockets they'd then bid our Hadean hellscape goodbye for the frozen calm of outer space.

A second atmosphere soon arose in place of the first one, an atmosphere summoned from the very ground itself. Just as there's oxygen dissolved in blood and carbon dioxide in champagne, magma (underground lava) has gases dissolved in it. And just as carbon dioxide whooshes out of open bottles, this magma would vent those gases when it reached a crack in Earth's surface and the pressure on the magma suddenly dropped. These escaping gases probably (geologists bicker about this) consisted mostly of water vapor and carbon dioxide, but other gases poured out as well. If you've ever stood near a volcanic rift, you can probably guess some of them, like the sulfurous compounds that give rotten eggs and gunpowder their bouquets (hydrogen sulfide and sulfur dioxide, respectively). The vents also released a faint haze of hot metallic vapors, including gold and mercury atoms. All fun stuff to breathe.

Some geologists have argued that all these gases came rushing out of the ground at once—the so-called Big Burp. Sadly for the juvenile-minded everywhere, the Big Burp probably never happened: Earth seems to have hoarded her gases instead, letting them seep out. Not that a gradual release made things more hospitable. The steam still would have scalded your skin, the sulfur gases still would have raked your nose, the acids and ammonia still would have shredded your lungs. The pressures involved weren't pleasant, either. Magma back then was far fizzier, having a much higher concentration of gases, and billions of tons of gases got released each day. The resulting air pressure, perhaps one hundred times modern levels, would have imploded your skull and crushed you into a much more spherical version of yourself. The slightest breeze in air this dense would have bowled you over and sent you tumbling into a pool of magma.

This second atmosphere isn't the same ocean of air we live under today, for a few reasons. The major component of air back then, steam, eventually condensed out as rain and began to pool on the ground in lakes and seas. The formation of seas and lakes also had knock-on effects, because the second-most-common ingredient in air then, carbon dioxide, dissolves readily in water and reacts with minerals there to form solid precipitates. This removes the CO_2 from circulation.

Another reason we don't live under thousands of pounds of pressure today is that stray asteroids (and/or comets) kept slamming into us and blasting that early air back into space. Not every impact was a disaster, mind you. Smaller asteroids probably even added gases to our atmosphere, from vapors trapped inside them. But every time our planet landed a big one, it learned a hard lesson about conservation of energy. Much of the kinetic energy of the asteroid's motion got transformed into heat when it struck home, heat that boiled atmospheric gases away into space. The rest of the kinetic energy created a gigantic shock wave that swept even more air up and away. Some geologists argue that impacts like these stripped our atmosphere bare multiple times, completely denuding us. If that's true, then instead of referring to our volcano-born atmosphere as *the* second atmosphere, we should probably talk about atmospheres 2a, 2b, 2c, and so on. Each one had the same mix of gases, but asteroids and comets kept evacuating them.

One of these evacuations deserves special attention, since it also created our moon. It's actually a strange moon, ours. Every other mooned planet we know of has mere gnats circling it, bodies far less massive; we have an albatross, a body one-quarter our size. To explain that anomaly, astronomers in the twentieth century devised several theories. Some argued that our moon formed independently, as a separate planet, and that we snared it in our gravitational mitt one day as it tried to slip past. Others, taking up a suggestion by Charles Darwin's son, astronomer George Darwin, argued that the moon somehow fissioned off from Earth in its not-quite-solid days, like a daughter cell budding off from a parent.

Moon rocks gathered in 1969 finally settled the issue, pointing to a combination of these theories. Among other clues, moon rocks had fewer volatile gases trapped inside them than Earth rocks, implying that something had boiled the lunar gases off. Boiling away all the gases inside a body as big as the moon, however, would have required massive amounts of heat, which suggests an unthinkably large collision: when astronomers ran the numbers, they found that this hypothetical impactor—now called Theia—was roughly as big as Mars. Theia likely formed as a separate planet at a different point on Earth's orbit, a few months "behind" or "ahead" of us as we circled the sun. But gravity, that eternal meddler, couldn't abide two planets circling in the same vicinity, and within about fifty million years of their formation decided to bang them together like rocks. Astronomers call it the Big Thwack.

If you're thinking here of the asteroid impact that wiped out the dinosaurs, think again. That strike produced pillars of fire right out of Exodus, sure, and threw up enough dust to dim the sun for several years. But Earth as a whole barely flinched: even after the sparrow cracked the windshield, the car kept going. Theia was more like plowing into a moose—major structural damage. The collision not only ejected our atmosphere, it may well have boiled our frickin' oceans and vaporized whole continents of rock. It also plowed deep into our mantle and left Earth lopsided, as if someone had punched a desk globe. Theia itself vaporized, and most of the gaseous remains began to stream into space and swirl around us, creating our very own celestial ring. But unlike the rings of Saturn, which are mostly ice and rocks, this piping-hot gas ring eventually cooled and coalesced into our moon.

In the long run Theia's impact gave us all sorts of poetic things, like full moons and even spring and autumn, since it nudged Earth sideways off its up-and-down axis and allowed for the variable sunlight that gives rise to seasons. In the short term, though, Theia managed to make our planet even less hospitable than before. In particular, we got an atmosphere even hotter and nastier than the volcanic atmosphere that preceded it. This atmosphere, which lasted for perhaps a thousand years, consisted mostly of scorching hot silicon (think vaporized sand) punctured by iron "rain." It would have had a salty tang as well, from fumes of sodium chloride; imagine every breath tasting like a salt lick.

Our new moon watched over all this from a height of just 15,000 miles; it loomed in the sky a dozen times larger than the sun does today. And its fiery, molten state would have left it glowing like a blackened, bloodshot eye. There might still have been poetry in this moon, but it was more Dante than Frost.

Eventually the solar system ran low on asteroids to shell Earth with (regularly, anyway). As a result, any gases that accumulated in our atmosphere could stick around instead of being blasted away into space. Just as important, proper volcanoes emerged. Volcanoes probably weren't common on the very early Earth, because the magma could vent its gases more easily through cracks in the semi-molten outer shell. But when space rocks stopped hitting us, our planet cooled and acquired a hard surface crust with far fewer cracks; magma began collecting in underground pools instead. Because that magma still contained dissolved gases, the pressure inside those pools often swelled to dangerous levels. Over time, the pressure would grow so high that lava and hot gases would burst through the overlying rock and scorch everything in their path.

Although magma today is far less fizzy than it used to be, this billennia-old cycle of pressure building up underground and then exploding upward continues today. In fact, it's exactly what happened with Mount Saint Helens in 1980.

———◄◦►———

Eddie's death depleted Truman. Days and weeks slipped by in lethargy, and he grew distracted and negligent at work. A tree he felled clobbered him on the head. He got his hand caught in a snowblower. He knocked himself unconscious falling off his porch and woke up in the snow in his tighty-whities. The campground suffered, too. Without Eddie to make up beds for guests, the cabins looked slovenly. Truman also flamed out as a cook. Eddie had never exactly earned a Michelin star at the café; among other delicacies, she served hot-dog sandwiches and hamburgers on soggy white bread. But Truman left diners with culinary PTSD. Lunch might be PB&O—that's peanut butter 'n' onions—while dinner might consist of "chicken back soup," a bird carcass boiled with twenty-five cloves of garlic. You almost wonder whether he was trying to drive customers away.

If Truman felt dispirited, though, the mountain looming over him grew more and more feisty each month, thanks to shifts in the continental plates beneath it. By 1980 the terms "continental drift" and "plate tectonics" were just starting to feel comfortable in the mouths of geologists—a situation that few of them could have predicted even fifteen years earlier. The first person to propose the theory of continental drift was German meteorologist Alfred Wegener, who spent the early 1900s puzzling over the fact that the edges of South America and Africa seemed to fit together like broken fragments of pottery. He noticed that the two continents shared similar fossil patterns as well, and after getting shot in the throat during

World War I, he decided to spend his convalescence writing a book in which he proposed that the continents rested on large plates that had somehow drifted over time.

To say that geologists didn't embrace Wegener's theory is a bit like saying that General Sherman didn't receive the warmest welcome in Atlanta. Geologists loathed plate tectonics, even got pleasure out of loathing it. But as more and more evidence trickled in through the 1940s and 1950s, the drifting of continental plates didn't seem so silly anymore. The balance finally tipped in the late 1960s, and in one of the most stunning reversals in science history, pretty much every geologist on Earth had accepted Wegener's ideas by 1980. The rout was so complete that nowadays we have a hard time appreciating the theory's importance. In the same way that the theory of natural selection shored up biology, plate tectonics took a hodgepodge of facts about earthquakes, mountains, volcanoes, and the atmosphere, and fused them together into one overarching schema.

Continental plates can sometimes shift all at once in dramatic fashion, jolts we call earthquakes. Much more commonly the plates slowly grind past one another moving at about the rate that fingernails grow. (Think about that next time you clip your nails: we're that much closer to the Big One.) When one plate slips beneath the other, a process called subduction, the friction of this grinding produces heat, which melts the lower plate and reduces it to magma. Some of this magma disappears into the bowels of Earth; but the lighter fraction of it actually climbs back upward through random cracks in the crust, swimming toward the surface. (That's why hunks of pumice, a volcanic rock, float when tossed into water, because pumice comes from material with such a low density.) The heat of grinding also liberates carbon dioxide from the melting plate, as well as lesser amounts of hydrogen sulfide, sulfur dioxide, and other gases, including trace amounts of nitrogen.

Meanwhile, as hot magma pushes upward through cracks in the

crust, water in the crust seeps downward through those same cracks. And here's where things get dangerous. One key fact about gases — it comes up over and over — is that they expand when they get warmer. Related to this, the gaseous version of a substance always takes up far more space than the liquid or solid version. So when that liquid water dribbling downward meets that magma bubbling upward, the water flashes into steam and expands with supernoval force, suddenly occupying 1,700 times more volume than before. Firefighters have special reason to fear this phenomenon: when they splash cold water onto hot, hissing fires, the burst of steam in an enclosed space can flash-burn them. So it goes with volcanoes. We ogle the orange lava pouring down the slopes, but it's gases that cause the explosions and that do most of the damage.

Around the world roughly six hundred volcanoes are active at any moment. Most of them lie along the famous Ring of Fire around the Pacific Ocean, which rests atop several unstable plates. In the case of Mount Saint Helens, the Juan de Fuca plate off Washington State is grinding away against the North American plate, and doing so roughly a hundred miles beneath the surface. This depth leaves a heavy cap of rock over the pools of magma and thereby prevents constant venting of noxious fumes. But when one of these deep pockets does pop, there's that much more shrapnel.

———<o>———

On March 20, 1980, a 4.0-magnitude tremor shook Mount Saint Helens. Feeling the ground shudder wasn't unusual in those parts, but unlike with previous earthquakes, this time the ground kept jiggling. In a typical five-year span, Mount Saint Helens might experience forty quakes. In the week after March 20, it got rocked by one hundred.

This put scientists in a delicate position. They didn't want to alarm the public about an eruption that might never happen. Just

five years earlier a doomsday prediction about the Mount Baker volcano north of Seattle had fizzled, making geologists look foolish.
Still, Mount Saint Helens wouldn't simmer down. On March 27 a
plume of smoke broke through the peak, rising 7,000 feet and staining the white snow there black. Shortly afterward state officials
closed all roads around Mount Saint Helens with barricades. More
controversially, they began forcibly evacuating residents. A young,
blond geologist named David Johnston explained the reasoning to
reporters: "This is like standing next to a dynamite keg. The fuse is
lit, but you don't know how long the fuse is."

One resident in the evacuation zone, however, decided that the
government didn't know what the hell it was talking about. Harry
Truman, just three miles distant from that beautiful, deadly cone,
dismissed the first quake as a mere avalanche. Even during the
onslaught of aftershocks, he refused to believe he was in danger.
He'd lived most of his life in the mountain's shadow, including his
best years, with Eddie, and he declared that "if the mountain did do
something, I'd rather go right here with it."

Word spread fast about The Man Who Wouldn't Leave the
Mountain, especially among reporters, who were finding the volcano beat frustrating. Aside from Johnston's "dynamite keg" quote,
reporters looking for hard facts couldn't get much out of geologists,
who hedged every prediction and then hedged the hedges. So most
stories went the other way, brushing over the science and emphasiz-
ing local color. As Truman used to say to photographers, you need a
goddamn human in the thing to keep people interested, and his
crotchety old man routine—he even wore socks with sandals
sometimes—played well in the media. For decades Truman had
actually shied away from publicity, still spooked that someone from
his bootlegging days might show up and smoke him. But hell, by
1980 that was half a century ago, and Truman discovered that he
liked telling old stories to fresh ears. Like that time he fought off a

bear with a rake, clad only in his underwear. Or the time he ginned up a local sasquatch legend by whittling two big wooden feet and leaving footprints in the snow. He also dusted off several World War I stories, and claimed that he had half a mind to strap on his old pilot's helmet and drop a bomb into the crater to shut it up.

Beyond telling stories, Truman bad-mouthed local officials every chance he got. "They say she's gonna blow again, but they're lying like horses trot," he sneered. He also boasted that he could judge the Richter strength of an earthquake faster than geologists could, just by watching how far the Rainier Beer sign in the café window swung side to side. Every major media outlet in the country came a-calling at some point, and every reporter knew to bring some Schenley whiskey as an offering. Truman soon had a cabinet full of bottles, and he hooted that the mountain could do its worst now — he could wait out anything.

Officials considered arresting Truman to enforce the evacuation order and clamp down on the hoopla. But what then? Throwing an old man in jail wouldn't exactly win the public over, and good luck finding a jury to convict. Better to let him spout off, even in the *New York Times* and *National Geographic*.

For the public, every week that passed without an explosion ratcheted up the tension and excitement. Yard signs in local communities read "St. Helens: keep your ash off my lawn." Given all the logging trails around the mountain, anyone with a decent map could circumvent the roadblocks, and people practically began having picnics on the slopes, thrilled at the prospect of seeing some real-life lava. Even Washington's governor, Dixy Lee Ray — who as a former scientist (a marine biologist) really should have known better — got swept up. "I've always said," she gushed, "that I hoped to live long enough to see one of our volcanoes erupt." At a low point in the hijinks, a Seattle film crew helicoptered in and filmed a beer commercial near the peak.

All the while, gases kept building up inside Mount Saint Helens. Planes circling overhead detected stronger and stronger whiffs of sulfur dioxide (the gunpowder smell), which meant that sulfur-rich magma was rising toward the surface. (Flying above the mountain was technically illegal at this point, but so many planes violated the ban that one pilot compared the swarm to an old-fashioned dogfight.) More ominously, in mid-April geologists noticed a tumor on the mountain's north side, a huge blister of bulging rock. No one knew how fast the bulge was growing, so planes went up with surveying equipment that had lasers capable of detecting a rise of even a few millimeters per day. A few millimeters doesn't sound that dramatic, until you realize that you're talking about lifting an entire mountain face. Turns out they didn't need such fine resolution anyway: the bulge was growing five feet taller every day. Government geologists had seen enough bad signs that on May 9 they pulled back their observation towers to a distance of six miles, well beyond the calculated danger zone. Truman chuckled at the scaredy-cats.

At least he did so publicly; privately, the pressure was building up inside him, too. Truman had always argued that the huge fir trees that stood between him and the volcano would buffer and protect him, but as the weeks wore on, friends could see his conviction faltering. He was more alone than ever, and although reporters laughed at his claims of wearing spurs to bed, to keep him from being bucked onto the floor, he was only half joking. Some nights the tremors would rattle the café dishes several times an hour, keeping him on edge. On the nights he did manage to fall asleep, he might wake up to find his bed pushed across the room. He didn't want to abandon his and Eddie's home, but that didn't mean he wanted to live each night in terror.

Sensing this, friends and officials made one last effort in mid-May to coax the wildcat down the mountain. Truman said no. He did so in part because by that point he was receiving sacks full of mail from people inspired by his courage. Marriage proposals trickled in, too. ("Now, why would some eighteen-year-old chick want to marry an old [*beep*] like me?" he marveled.) He also got a letter from Dixy Lee Ray praising his steadfastness, which he practically waved over his head in excitement. It's hard to say whether the fame and adulation reinforced what he wanted to do anyway, or whether the pressure of being famous forced him into decisions he wouldn't otherwise have made. Either way, he wasn't going anywhere.

Giving up on Truman, state officials decided to clear everything else off the mountain. After three days of seismic peace, they even lifted the blockade for local cabin owners on May 17, allowing them to dash up the roads in empty trucks and pack up chairs, tables, toasters, cameras, and whatever else they could grab. Officials gave permission for a second run the next morning. The mountain had other plans.

———◄○►———

Despite all the hedging they did, Mount Saint Helens still managed to embarrass scientists. It wasn't so much missing the *when* that bothered them. Eruptions are impossible to predict, so there was no shame in not having Sunday, May 18, at 8:32:11 a.m. in the office pool. But geologists had all agreed on *what* would happen when the trigger came—that the mountain's cone would channel all the force upward and discharge the gas and smoke into the sky. Not quite. Instead, the blister on the north face popped and, after a brief collapse inward, all the nasty stuff started gushing out sideways, the largest landslide in recorded history. In particular the supercharged gases came screaming down the slopes, mixing with the ash and smoke and vaporizing everything in their path.

You could mark the devastation in a series of concentric circles. People two hundred miles away heard the boom. A hundred miles

The eruption of Mount Saint Helens on Sunday, May 18, 1980. (Photo courtesy U.S. Forest Service)

distant, windows rattled in their frames and heirloom china toppled over inside cabinets. Eighty-five miles east, in Yakima, mudballs plopped down from the sky, and the sky itself grew black enough to trigger automatic streetlights at 9:30 a.m. Forty-five miles away, the temperature in certain streams rose past 90°F, forcing salmon to jump onto the banks.

Oddly, the closer you got to the mountain, the less likely you were to hear the boom. The thick sand and debris absorbed much of the noise in the immediate vicinity. Sound also tends to rise in warm air close to the ground, so the noise sailed over the heads of most people nearby. (More on this phenomenon later.) The lack of a roar didn't mean that things were peaceful within the blast zone, however. Every tree within fifteen miles got flattened, as if someone had taken a giant comb and parted Mount Saint Helens's hair. Logging trucks weighing several tons flipped onto their sides like matchbox cars. People were pelted with flaming rubble and had their hair singed off. A man and a woman caught inside this zone, who survived by jumping into the hollow of an uprooted tree, made a pact that they would get married if they survived. They did, and did.

Within about ten miles almost everything that could have died did. This included seven thousand large mammals (elk, deer, bears), as well as David Johnston, the blond geologist who'd compared Mount Saint Helens to a dynamite keg. Young and fit (he ran marathons, which was odd back then), Johnston had been volunteering for shifts in the closest observation towers, since he figured he had a better chance of escaping than older colleagues. But he wasn't supposed to work on May 18 at all. He'd swapped shifts that day with his boss, who wanted to spend time with a friend visiting from Germany. When Mount Saint Helens blew, Johnston had just enough time to radio in to headquarters, shouting, "This is it!" Colleagues said he sounded overjoyed.

Johnston's body was never found, but dozens of others did turn up within this zone. Coroners determined that most of them had died from inhaling ash, because as soon as the ash hit the saliva in their mouths it thickened into putty and blocked their airways. After they died, their bodies broiled in the heat and their inner organs dehydrated into jerky. A few bodies were so gamy and saturated with ash that they dulled the blades of scalpels.

Within five miles, rescuers didn't even bother looking for bodies. The smell of sulfur would have been overwhelming here, a flashback to early Earth. And like early Earth, the landscape was devoid of features. When Jimmy Carter flew over a few days later, he

Mailboxes nearly buried in hot ash from the eruption of Mount Saint Helens. (Photo courtesy U.S. Geological Survey)

marveled, "It makes the surface of the moon look like a golf course." The only landmark now was the reborn Spirit Lake. Like everything else, the original Spirit Lake had been buried beneath dozens of yards of smoldering ash. But the water managed to force its way up through the debris and re-form into a new lake sixty yards higher. Whereas the original had been cool, clear, and inviting, with a visibility of thirty feet, Spirit Lake II looked like a steaming brown cesspool, with a visibility of inches.

Finally, within three miles, sat Harry Truman. Suffice it to say we'll never know for certain what happened to him, beyond the obvious fact that he died. But we can reconstruct his last moments, based on the victims of other volcanoes and the laws of chemistry.

First thing, his clothing would have gone up like flash paper in the heat—his jeans and sweater and socks all incinerating in an instant. The enamel on his teeth would have cracked, and his once-moist lungs would have blackened into brittle carbon. At the same time, he probably died without suffering. Many bodies recovered after the Mount

Vesuvius eruption in AD 79, especially those of people who took shelter in chambers along the beach, showed no signs of fear or struggle;
none were shielding their faces with their hands, for instance, a common sign of agony. This suggests that heat shock probably killed them
in less than a second—too fast for their reflexes to register any pain.

The heat of Vesuvius also caused many victims' muscles to contract after they died, furling their toes beneath their feet and curling
their arms upward toward their chests—the so-called pugilistic
pose. Their bodies were then buried in ash, which hardened around
them like a death mask. Bizarrely, as they decayed, a human-shaped
hole was left behind. By filling these holes with plaster, archaeolo-
gists created the casts of bodies that tourists now see at Pompeii.

So that's one possible end: a Harry Truman–shaped cavity buried several yards beneath the volcanic crust. But there's another,
more macabre possibility. Vesuvius devastated other villages
besides Pompeii, and in those places—which received a direct blast
of heat, much like Truman's campground—several people vaporized on contact.

The very few studies on body vaporization have broken down the
process into three stages: vaporizing water, vaporizing viscera,
vaporizing bones. Vaporizing water requires two steps. First you
have to raise the water inside you from body temperature (around
98.6°F) to boiling temperature (212°F). The water content of human
beings actually changes as we age, dropping from 75 percent in
squishy newborns to less than 60 percent in (somewhat literally)
crusty old folk. Given Truman's age and weight, he probably carried
100 pounds of water inside him. Water absorbs heat quite well, so
raising that much water to 212°F requires lots of energy, around
2,900 food calories. (By comparison, it takes 305 calories to heat up
100 pounds of iron by that same amount; with gold it takes 88.) And
all we have so far is hot water. Creating actual steam requires far
more energy still. That's because water molecules have a strong

attraction to their neighbors; they therefore hesitate to leave those neighbors behind and fly up into the air to become gas. As a result it takes an additional 24,000 calories to create 100 pounds of steam.

As for vaporizing viscera, there are hundreds of organs and tissues involved here, all of which have different properties and heat-absorption profiles. Rather than tabulate those differences, some scientists use a proxy: dried pork, since humans and pigs have similar viscera. An average human has (after you subtract the water) roughly 25 pounds of organs and gristle and fat, and given that dried pork has about 230 calories per hundred grams, it probably takes around 27,000 more calories to break down all the viscera molecules.

Based on the state of victims near Vesuvius, we know that water and viscera boil away easily enough under the onslaught of volcanic heat. Vaporizing the 25-pound human skeleton is harder, since the main mineral in bone (calcium hydroxyapatite, $Ca_{10}(PO_4)_6(OH)_2$) has a very high boiling point. As a result, Truman's skeleton probably remained intact. That doesn't mean it emerged unscathed. Above about 900°F, bone turns pale yellow, then reddish brown, then black; it also melts slightly as its molecules rearrange themselves. This rearrangement weakens bone, making it crumbly. Some victims at Vesuvius also had the tops of their heads blown off. The hot volcanic gases seem to have boiled their brains, and because the brain vapors had to escape somewhere, they erupted out the top of their skulls like a mini–Mount Saint Helens.

Altogether, boiling the water and vaporizing the guts and skeleton of an old bastard like Harry Truman would have taken something like 75,000 calories,* a good month's worth of food intake. That's even more impressive considering that you'd have to deliver all that

---

*This and all upcoming asterisks refer to the Notes and Miscellanea section, which begins on page 329 and goes into more detail on various interesting points.

energy in one blow. Few things outside of atom bombs can vaporize a human body between heartbeats, but volcanoes belong to this club.

The last moments of Harry Truman's life would have gone something like this. That close to the blast, he likely heard nothing, though the ground would have rumbled beneath him and thrown him off balance. If he happened to look up, he would have seen something sublime, in the old poetic sense of both awful and awe-full. The mountain face he'd spent most of his life marveling over would have collapsed like a soufflé and then rebounded, with all the gases inside making their break for freedom at once. Given the speed of the black cloud that emerged (up to 350 miles an hour), Truman could have watched it careen down the mountain for maybe a minute—a plume one hundred stories tall and ten miles wide. Its intense heat would have boiled away any nearby snow and twisted those 250-foot fir trees like scraps of plastic in a campfire. When the front roared into Truman's campsite it would have blistered the paint on his pink Cadillac and perhaps burst the tires as the air inside them expanded. Every bottle of Schenley in the cabinet would have exploded like a Molotov cocktail and all the panther pee would have gone up in blue flames. Truman's clothes would have flared and disappeared, and then Truman himself would have sublimed in the scientific sense—transformed from solid to spirit almost instantly. And with a final hiss, he would have risen up into the air, joining the wider atmosphere.

Like Caesar's dying gasp, then, there's a decent chance you inhaled a bit of Harry Truman with your most recent breath.

———◇———

It took Mount Saint Helens two thousand years to build up its beautiful cone and about two seconds to squander it. It quickly shrank from 9,700 feet to 8,400 feet, shedding 400 million tons of weight in the process. Its plume of black smoke snaked sixteen miles high and created its own lightning as it rose. And the dust it spewed swept across the entire United States and Atlantic Ocean, eventually circling the world and washing over the mountain again from the west seventeen days later. Overall the eruption released an amount of energy equivalent to 27,000 Hiroshima bombs, roughly one per second over its nine-hour eruption.

With all that in mind, it's worth noting that Mount Saint Helens was actually small beer as far as eruptions go. Although it vaporized a full cubic mile of rock, that's only 8 percent of what Krakatoa ejected in 1883 and 3 percent of what Tambora did in 1815. Tambora also decreased sunlight worldwide by 15 percent, disrupted the mighty Asian monsoons, and caused the infamous Year Without a Summer in 1816, when temperatures dropped so much that snow fell in New England in summertime. And Tambora itself would tremble before the truly epic outbursts in history, like the Yellowstone eruption 2.1 million years ago that launched 585 cubic miles of Wyoming into the stratosphere. (This megavolcano will likely have an encore someday and will bury much of the continental United States in ash.)

At the end of Earth's Hadean, volcanic era, our planet had already experienced two distinct atmospheres—a first, wispy one consisting mainly of hydrogen and helium; and a second, much harsher one consisting of the dragon's breath of volcanoes. That latter atmosphere has of course long since passed—your lungs don't sizzle and shriek every time you breathe nowadays. But we can still catch glimpses of it—deadly glimpses—in certain places, the most spectacular example being a strange eruption that took place near Lake Nyos in Cameroon on August 21, 1986.

Interlude: The Exploding Lake

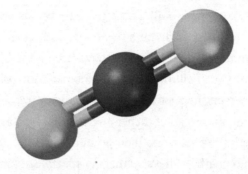

Carbon dioxide (CO_2)—currently 400 parts per million in the air (and rising); you inhale 500 quadrillion molecules every time you breathe

At first it sounded like another gang fight. The volcanic highlands of western Cameroon had always been sparsely populated, in part because many local tribes considered the region haunted. (One legend held that a vindictive spirit rose from nearby Lake Nyos at night and terrorized people living in the valleys below.) But by the 1980s the rich soil there, ideal for yams and beans, had lured farmers from several different ethnic groups to the region. And as the population swelled, so had gang violence and shootings. Many residents therefore dismissed the distant *pop-pop-popping* noises that night in August 1986 as nothing more than the usual gunfire.

Other noises that evening weren't so easy to explain—low rumbles, a strange gurgling. A blind woman felt the ground beneath her tremble several times, like someone shuddering. Most people nevertheless ignored the omens and went to bed. But shortly after sunset, the Earth really did summon something evil from the bottom of Lake Nyos.

Around nine o'clock, the lake began belching up huge bubbles of carbon dioxide. The bubbles carried with them water from the iron-rich bottom of the lake, giving them a red hue, and they began to *pop, pop* as they breached the surface. Five hundred million pounds of CO_2 escaped in all, and a fountain of gas and water shot 250 feet into the air, roaring for a full twenty seconds.

In other circumstances this might have been a thrilling sight—a blood-colored Old Faithful. But because carbon dioxide is so heavy—50 percent heavier than air—this huge mass of gas sank toward the ground instead of dispersing, eventually forming a white cloud 150 feet tall. This cloud then poured down the slopes around Lake Nyos into nearby valleys, gathering speed as it tumbled. These valleys were cooler, and as the water vapor in the cloud condensed out, the mass of gas became invisible—perfect for a nocturnal hunt.

Traveling at forty miles an hour, the miasma engulfed several villages within minutes—Cha, Subum, Fang, Mashi. Whole families were smothered as they slept in their mud huts; others were felled as they tended their dinner fires. Some people passed out so quickly they couldn't get their hands down to break their falls, and they suffered fractures upon landing. They then suffocated as carbon dioxide displaced the oxygen in their lungs. Three victims here, four in the next hut over, six more down the road.

By the next morning the water level of Lake Nyos had dropped a full three feet and its once azure surface reportedly looked like an oily tiger skin—orange with black streaks. Miraculously, a few people in the more distant villages survived, but they woke up with headaches, nausea, and diarrhea; many also had pressure sores from lying motionless on the same spot for hours. Worse, they were confronted with a holocaust: 1,746 people dead. The casualties continued to mount over the next few days as devastated parents and grandparents committed suicide and expectant mothers miscarried. Over 6,000 cattle died, too, along with nearly every mouse, bird, and insect in the region. Survivors remember an eerie stillness for days afterward. There weren't even flies around to snack on the dead bodies.

Cattle lie dead near mud huts after the mysterious "eruption" of Lake Nyos on August 21, 1986. (Photo courtesy U.S. Geological Survey)

Local diviners declared that the evil spirits in the lake had stirred to life again. Less superstitious (if more paranoid) types blamed the deaths on chemical warfare by their enemies or a clandestine neutron bomb test. Scientists, meanwhile, pointed to a more prosaic explanation. As the first sketchy news reports emerged, geologists from across the globe—few of whom had heard of Nyos—began pouring into Cameroon, eager to get a look at this "exploding lake." They determined pretty quickly that Nyos was resting atop an active volcano. (An eruption had actually formed this crater lake in the first place just four hundred years before.) Because volcanoes often release carbon dioxide, the crater was clearly the source of the gas.

Beyond that, however, geologists agreed on very little. There was especially heated disagreement about what had triggered the release of gas. Some argued that all the gas burst out during a single eruption; they pictured Nyos as a miniature Mount Saint Helens. Others held that no, an eruption had nothing to do with it. Instead, carbon dioxide had been building up at the bottom of Nyos for centuries, slowly leaking from a volcanic vent. The weight of the overlying water had kept the gas in place for a while, but a landslide or heavy rain had disturbed the lake bottom and allowed it to escape. Still others agreed about the slow buildup of CO_2, but disputed the idea of an external trigger. They claimed that the pocket of gas had simply grown too heavy for the overlying water one unlucky day and had finally pushed its way to the surface, like bubbles effervescing out of champagne.

Confusingly, partisans of both the single-eruption theory and the slow-buildup theory could point to evidence supporting their argument. Some survivors recalled smelling rotten eggs and sulfurous gunpowder before passing out, smells associated with volcanic eruptions. There's also no solid evidence of a landslide or strange weather pattern that might have disturbed the lake bottom. Moreover, the outburst did release loads of heat: the temperature of Lake Nyos jumped from 73°F to 86°F right after the event, and the fever persisted for weeks. Then again, a seismograph a hundred miles away recorded no geological activity that day, which makes an eruption seem unlikely. And however sincere, the testimony of survivors remains suspect, because CO_2 poisoning can cause confusion and mental fogginess. Some scientists suggested that carbon dioxide can also cause olfactory hallucinations, or that aid workers might have inadvertently planted ideas in people's heads by asking about those sulfurous gases in the days after the event.

The two camps eventually divided along nationalistic lines, with French, Italian, and Swiss geologists supporting the volcano theory and American, German, and Japanese geologists supporting the slow-buildup theory. One onlooker lamented that "non-experts can do little more than wave the national flag they fancy." (I should note that most experts today do favor the slow-buildup scenario—but as an American, I suppose I would say that...)

The squabbling not only makes scientists look petty, it might jeopardize the future safety of the region. If another eruption takes place, the villages around Lake Nyos are more or less doomed to suffer a repeat disaster, since scientists can't really predict, much less stop, eruptions. But they might be able to stop a slow buildup of carbon dioxide, especially if it takes years for enough gas to accumulate.

Perhaps for that reason—at least it gives them hope—the people of Cameroon have thrown in their lot with the slow-buildup theory and have spent the past twenty years trying to defuse their exploding lake. At first government officials debated several ways of doing this, including dropping bombs into Nyos. They finally decided to float rafts on the surface and snake a 666-foot polyethelene pipe toward the bottom for controlled ventings. These ventings are certainly spectacular: water goes shooting 150 feet into the air sometimes. Still, no one knows whether they'll do any good. Heck, mucking around down there might even backfire and trigger another outburst. Even today scientists warn visitors not to splash around too much when they wade into Nyos, lest they stir up something evil.*

Survivors still see signs of the 1986 disaster here and there. Water from the lake bottom feeds several nearby springs, and every so often they find a hawk or a mouse slumped over beside one, suffocated by a sudden belch of CO_2. Many locals claim that the ghosts of the dead still haunt the edges of Lake Nyos as well. "You hear them talking sometimes," one man insisted.

Exactly what the ghosts tell him, he didn't say. But their deaths, and the memories of that wretched August night, do bear witness to something important, something all of us should appreciate. That night provided an awful glimpse of what Earth once was, a place where megabubbles of poisonous gas broke free all the time and stalked the landscape like supernatural terrors. Above all, the Nyos disaster should remind us just how lucky our planet was to have escaped that Hadean state.

So how did we get from there to here, from toxic air to a comfortable, breathable atmosphere? The answer has several parts, but it depends in large measure on the rise of the next major gas in the history of Earth, nitrogen.

CHAPTER TWO

The Devil in the Air

Nitrogen (N_2)—currently 78 percent of the air (780,000 parts per million); you inhale nine sextillion molecules every time you breathe

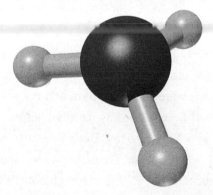

Ammonia (NH_3)—currently 0.00001 parts per million in the air; you inhale 100 billion molecules every time you breathe

For several hundred million years after its birth, Earth was pretty inhospitable. Even if you could have found a place to stand without burning your tootsies, you wouldn't have been able to breathe at all, thanks to the gases that volcanoes were pumping out. Amazingly, though, however noxious in the short term, volcanoes eventually redeemed our air by pumping out gases rich in nitrogen.

Three elements—oxygen, hydrogen, carbon—make up 93 percent of human beings' body weight and a similar percentage in other life-forms. Cells also require dozens of other elements to function, right down to obscurities like molybdenum. Unless something goes really wrong, animals and plants can cull most of those elements from the environment with a minimum of fuss.

The big exception is nitrogen. It's the fourth-most-abundant element inside us, making up a full 3 percent of our body weight. And it's by far the most common element in the air, making up four of every five molecules we breathe, day in, day out. So getting it into our cells should be a snap, right? Nope. Despite this abundance, most creatures have to fight and scrape for every atom of nitrogen they can get. That's because the cells of most creatures, including humans, can't utilize nitrogen in its gaseous form. Nitrogen has to undergo conversion into another form first. And for the first few billion years of Earth's history, only a few special microbes had mastered the necessary sleight of hand.

In the early 1900s, however, *Homo sapiens* became the first non-bacterium to join the nitrogen-making pantheon. The two men responsible for this were both German, and both worked as industrial chemists. Both were hailed as national heroes for their discoveries, and both won Nobel Prizes. Both were also later condemned as international war criminals. But however much people hated them, they did succeed in coaxing element 7 out of the sky and into our bodies. You simply can't tell the story of air without the chemical magic of Fritz Haber and Carl Bosch.

Chemist Fritz Haber, one of the most Faustian figures in science history.

The alchemy of air started with an insult. Fritz Haber was born into a middle-class German Jewish family in 1868, and despite an obvious talent for science, he ended up drifting between several different industries as a young man—dye manufacturing, alcohol production, cellulose harvesting, molasses production—without distinguishing himself in any of them. Finally, in 1905 an Austrian company asked Haber—by then a balding fellow with a mustache and pince-nez glasses—to investigate a new way to manufacture ammonia gas (NH_3).

The idea seemed straightforward. There's plenty of nitrogen gas in the air (N_2), and you can get hydrogen gas (H_2) by splitting water molecules with electricity. To make ammonia, then, simply mix and heat the gases: $N_2 + 3H_2 \rightarrow 2NH_3$. Voilà. Except Haber ran into a heckuva catch-22. It took enormous heat to crack the nitrogen molecules in half so they could react; yet that same heat tended to destroy the product of the reaction, the fragile ammonia molecules. Haber spent months going in circles before finally issuing a report that the process was futile.

The report would have languished in obscurity — negative results win no prizes — if not for the vanity of a plump chemist named Walther Nernst. Nernst had everything Haber coveted. He worked in Berlin, the hub of German life, and he'd made a fortune by inventing a new type of electric lightbulb. Most important, Nernst had earned scientific prestige by discovering a new law of nature, the Third Law of Thermodynamics. Nernst's work in thermodynamics also allowed chemists to do something unprecedented: examine any reaction — like the conversion of nitrogen into ammonia — and estimate the yield at different temperatures and pressures. This was a huge shortcut. Rather than grope blindly, chemists could finally *predict* the optimum conditions for reactions.

Still, chemists had to confirm those predictions in the lab, and here's where the conflict arose. Because when Nernst examined the data in Haber's report, he declared that the yields for ammonia were impossible — 50 percent too high, according to his predictions.

Haber swooned upon hearing this. He was already a high-strung sort — he had a weak heart and tended to suffer nervous breakdowns. Now Nernst was threatening to destroy the one thing he had going for himself, his reputation as a solid experimentalist. Haber carefully redid his experiments and published new data more in line with Nernst's predictions. But the numbers remained stubbornly higher, and when Nernst ran into Haber at a conference in May 1907, he dressed down his younger colleague in front of everyone.

Honestly, this was a stupid dispute. Both men agreed that the industrial production of ammonia via nitrogen gas was impossible; they just disagreed over the exact degree of impossibility. But Nernst was a petty man, and Haber—who had a chivalrous streak—could not let this insult to his honor stand. Contradicting everything he'd said before, Haber now decided to prove that you could make ammonia from nitrogen gas after all. Not only could he rub Nernst's fat nose in it if he succeeded, he could perhaps patent the process and grow rich. Best of all, unlocking nitrogen would make Haber a hero throughout Germany, because doing so would provide Germany with the one thing it lacked to become a world power—a steady supply of fertilizer.

Ammonia is a gateway to fertilizer. That's not simply because ammonia contains nitrogen; it's more that ammonia contains nitrogen in a form that plants can exploit. To see the distinction here, you need to know something about the bonds that hold atoms together within molecules. Most molecules consist solely of single bonds (X-Y) or double bonds (X=Y). Nitrogen gas, however, invests in a triple bond (N≡N), one of the strongest and least breakable bonds in nature. (The sum total of the triple bonds in just one ounce of nitrogen stores enough energy to lift a million-pound dumbbell fifteen inches off the floor.) The strength of this bond explains why nitrogen dominates our air today. As mentioned in the last chapter, nitrogen is a mere trace component of most volcanic eruptions, far less common than other gases that spew forth. But whereas most volcanic gases disappear over time—either because they react with one another or because ultraviolet light splits them apart—nitrogen's triple bond resists all degradation. This allows the teeny-tiny percentage of nitrogen in any one eruption to build up over time. (Additional N_2 formed when the ammonia in volcanic fumes split apart.) In other words, nitrogen dominates our air today because it has outlasted everything else that volcanoes spat out.*

The big-picture consequence of this was that Earth's atmosphere got remade again. The second atmosphere, remember, was teeming with harsh volcanic exhaust. But by about two billion years ago, enough of those harsh gases had degraded—and enough nitrogen had accumulated—to count as something new (our planet's third distinct atmosphere, for those counting). Crucially for life, this third, nitrogen-rich atmosphere was far more tranquil and soothing, since nitrogen doesn't attack biological molecules the way that other gases do.

But in one sense aerial nitrogen is *too* tranquil, too passive. Our bodies need a fair amount of this element. Every scrap of protein inside you has a string of nitrogen atoms in its backbone, and every one of your 3 billion DNA bases in every one of your 30 trillion cells contains several nitrogen atoms as well. But when it comes time to stock our cells with nitrogen, the triple bond in N_2 renders it nonreactive. It's a cruel irony, really. We live under a sea of nitrogen gas: several *tons* of it hover over your head at all times, suspended between the ground and space. Yet we can't use any of it. It's like dying of thirst in the middle of the ocean.

So how does nitrogen get out of the air and into our bodies? Something has to "fix" it—break the triple bond and convert it into a less aloof form. Lightning can fix a little nitrogen by creating nitrogen-oxygen compounds in the air. But the vast majority of fixed nitrogen comes from bacteria that contain a special enzyme called nitrogenase. Enzymes are biological structures that allow unusual reactions to take place, and the business end of the nitrogenase molecule is a nub of iron and sulfur and molybdenum atoms. Like a tiny Jaws of Life, these elements tear the triple bond apart rung by rung. This takes tremendous energy and actually produces a fair amount of collateral damage: 16 water molecules get sacrificed each time. But in the end nitrogenase splits the $N{\equiv}N$, and before the nitrogens can jump together again, the enzyme welds on some hydrogen atoms. This creates (*ta-da!*) ammonia, which contains only single bonds and can therefore be converted into proteins or DNA fairly easily.

Nitrogen-fixing bacteria live in the roots of certain plants, where they barter their ammonia for other nutrients—symbiosis at its best. Other nitrogen fixers do freelance work in the soil, whereupon plants slurp it up. Organisms such as animals and fungi get their nitrogen fix by eating these plants or feeding on decayed plant remains. (This includes carnivores atop the food chain, who eat the herbivores that ate the plants. Even carnivorous plants, like Venus flytraps, chomp on bugs largely to get at their nitrogen.) Ultimately, then, the nitrogen inside virtually every organism on Earth comes from these bacteria. Without them not a single plant or animal would exist. Not one. And in most ecosystems, the amount of nitrogen in the soil sets a maximum on how much life the ecosystem can support.

That said, farmers developed a few tricks over the millennia to sidestep that limit on the soil. They began rotating crops to incorporate plants like soybeans, whose roots contain oodles of nitrogenase bacteria. They also started spreading urine and manure on their fields, waste products that get broken down into fixed nitrogen. Some clever folks even enhanced their manure by composting it with blood or other rotting matter. These heaps of compost looked like giant loaves of brown bread and radiated heat; farmers knew they were ready to spread when the mixture tasted spicy.

Still, domestic animals can poop and pee only so much, and by the 1800s most industrialized nations had to reach beyond their borders to meet their nitrogen needs. Great Britain leaned heavily on India, where poor, low-caste laborers processed manure for export by stamping cow pies and urine into a mash with their feet. Other European countries began mining vast deposits of guano (bird poop) from various islands around the world. Soon enough the guano trade on one set of islands—the Chincha Islands, off Peru*— grew so lucrative that several South American countries actually went to war over these piles of bird shit. Guano also compelled the

United States to jump into colonialism in a big way. In 1856 Congress passed the Guano Islands Act, which deputized any U.S. citizen to claim any unoccupied island anywhere in the world, provided it had a few smears of guano on it. This provided legal cover for the United States to seize close to one hundred specks in the Caribbean and Pacific. Many were godforsaken rocks that never amounted to anything, but several, including Johnston and Midway Islands, served as vital military bases during World War II. The Allies might never have defeated Japan in the Pacific if not for our guano greed a century before.*

Germany, meanwhile, missed out on the good guano gettin'. Unlike its rivals in Europe, Germany had only recently (during Haber's boyhood) put aside its long-standing tribal squabbles and united as a single country. As a result it had missed out on the great colonial land grabs in Asia and America and had few colonies to exploit for cheap guano. To compound the problem, Germany had poor native soil and needed that nitrogen fertilizer badly. By the early 1900s it was importing 900,000 tons each year.

However poor the soil, something in the German water sprouted scientific genius, and a German chemist in the 1840s came up with the radical idea of *artificial* nitrogen fertilizer, so-called chemical manure. Several decades would pass before people took the idea seriously, what with the South American guano flowing so freely. But by the 1890s, the fertilizer industry was facing a crisis, as miners had all but exhausted deposits in the Chinchas and elsewhere. Science alone seemed capable of staving off mass starvation. And here's where we finally circle back to Fritz Haber, a man ripe with talent and ambitious beyond all bounds of decency.

After Nernst humiliated him, Haber secured a grant from a German chemical company named BASF to explore a few possible technologies for fixing nitrogen. One involved creating literal lightning in a bottle—huge electric sparks inside vats of air, to fuse nitrogen and oxygen. Haber, however, preferred to focus on his old idea of fusing nitrogen and hydrogen, in part because he'd developed a new approach here—cranking up the pressure.

Both high temperatures and high pressures encourage gases to react more quickly. On a molecular level, raising the temperature of a gas causes its molecules to zip around at higher speeds. (In fact, that's essentially what temperature measures—molecular tempo.) The oomph from that extra speed allows molecules to break apart and recombine more easily, which encourages reactions. But Haber knew that adding extra heat would also destroy the ammonia, rendering the higher rate of reaction moot. That's why he focused on pressure. High pressure brings molecules into more intimate contact, giving them more opportunity to intertwine and swap atoms.

Several chemists had already tried to raise the pressure on nitrogen and hydrogen for this reason; one man, for instance, used a converted bicycle pump. But such equipment fell laughably short of what Haber proposed—pressures hundreds of times higher than atmospheric levels, pressures capable of crushing modern submarines. To achieve such pressures, Haber designed 30-inch tubes made of quartz and reinforced with iron jackets. This allowed him to lower the reaction temperature by a few hundred degrees and preserve more ammonia.

Beyond fiddling with temperatures and pressures, Haber focused on a third factor, a catalyst. Catalysts speed up reactions without getting consumed themselves; the platinum in your car's muffler that breaks down pollutants is an example. Haber knew of two metals, manganese and nickel, that boosted the nitrogen-hydrogen reaction, but they worked only above 1300°F, which fried the ammonia. So he scoured around for substitute catalysts, streaming these gases over dozens of different metals to see what happened. He finally hit upon osmium, element 76, a brittle metal once used to make light-bulbs. It lowered the necessary temperature to "only" 1100°F, which gave ammonia a fighting chance.

Using his nemesis Nernst's equations, Haber calculated that osmium, if used in combination with the high-pressure jackets, might boost the yield of ammonia to 8 percent, an acceptable result at last. But before he could lord his triumph over Nernst, he had to confirm that figure in the lab. So in July 1909—after several years of stomach pains, insomnia, and humiliation—Haber daisy-chained several quartz canisters together on a tabletop. He then flipped open a few high-pressure valves to let the N_2 and H_2 mix, and stared anxiously at the nozzle at the far end.

It took a while: even with osmium's encouragement, nitrogen breaks its bonds only reluctantly. But eventually a few milky drops of ammonia began to trickle out of the nozzle. The sight sent Haber racing through the halls of his department, shouting for everyone to "Look! Come look!" By the end of the run, they had a whole quarter of a teaspoon.

They eventually cranked that up into a real gusher—a cup of ammonia every two hours. But even that modest output persuaded BASF to purchase the technology and fast-track it. As he often did to celebrate a triumph, Haber threw his team an epic party. "When it was over," one assistant recalled, "we could only walk home in a straight line by following the streetcar tracks."

———◦———

Haber's discovery proved to be an inflection point in history—right up there with the first time a human being diverted water into an irrigation canal or smelted iron ore into tools. As people said back then, Haber had transformed the very air into bread.

Still, Haber's advance was as much theoretical as anything: he proved you could make ammonia (and therefore fertilizer) from nitrogen gas, but the output from his apparatus barely could have nourished your tomatoes, much less fed a nation like Germany. Scaling Haber's process up to make tons of ammonia at once would require a different genus of genius—the ability to turn promising ideas into real, working things. This was not a genius that most BASF executives possessed. They saw ammonia as just another chemical to add to their portfolio, a way to pad their profits a little. But the thirty-five-year-old engineer they put in charge of their new ammonia division, Carl Bosch, had a grander vision. He saw ammonia as potentially the most important—and lucrative—chemical of the new century, capable of transforming food production worldwide. As with most visions worth having, it was inspiring and dicey all at once.

Chemist Carl Bosch, mastermind of nitrogen-ammonia production.

Bosch decided to tackle each of the many subproblems with ammonia production independently. One issue was getting pure enough nitrogen, since regular air contains oxygen and other "impurities." For help here, Bosch turned to an unlikely source, the Guinness Brewing company. Fifteen years earlier Guinness had developed the most powerful refrigeration devices on the planet, so powerful they could liquefy air. (As with any substance, if you chill the gases in the air enough, they'll condense into puddles of liquid.) Bosch was more interested in the reverse process — taking cold liquid air and boiling it. Curiously, although liquid air contains many different substances mixed together, each substance within it boils off at a separate temperature when heated. Liquid nitrogen happens to boil at −320°F. So all Bosch had to do was liquefy some air with the Guinness refrigerators, warm the resulting pool of liquid to −319°F, and collect the nitrogen fumes. Every time you see a sack of fertilizer today, you can thank Guinness stout.

The second issue was the catalyst. Although effective at kick-starting the reaction, osmium would never work in industry: as an ore it makes gold look cheap and plentiful, and buying enough osmium to produce ammonia at the scales Bosch envisioned would have bankrupted the firm. Bosch needed a cheap substitute, and he brought the entire periodic table to bear on the problem, testing metal after metal after metal. In all, his team ran twenty thousand experiments before finally settling on aluminum oxide and calcium mixed with iron. Haber the scientist had sought perfection — *the* best catalyst. Bosch the engineer settled for a mongrel.

Pristine nitrogen and cut-rate catalysts meant nothing, however, if Bosch couldn't overcome the biggest obstacle, the enormous pressures involved. A professor in college once told me that the ideal piece of equipment for an experiment falls apart the moment you take the last data point: that means you wasted the least possible amount of time maintaining it. (Typical scientist.) Bosch's equipment had to run for months without fail, at temperatures hot enough to make iron glow and at pressures twenty times higher than in locomotive steam engines. When BASF executives first heard those figures, they gagged: one protested that an oven in his department running at a mere seven times atmospheric pressure—one-thirtieth of what was proposed—had exploded the day before. How could Bosch ever build a reaction vessel strong enough?

Bosch replied that he had no intention of building the vessel himself. Instead he turned to the Krupp armament company, makers of legendarily large cannons and field artillery. Intrigued by the challenge, Krupp engineers soon built him the chemistry equivalent of the Big Bertha: a series of eight-foot-tall, one-inch-thick steel vats. Bosch then jacketed the vessels in concrete to further protect against explosions. Good thing, because the first one burst after just three days of testing. But as one historian commented, "The work could not be allowed to stop because of a little shrapnel." Bosch's team rebuilt the vessels, lining them with a chemical coating to prevent the hot gases from corroding the insides, then invented tough new valves, pumps, and seals to withstand the high-pressure beatings.

Beyond introducing these new technologies, Bosch also helped introduce a new approach to doing science. Traditional science had always relied on individuals or small groups, with each person providing input into the entire process. Bosch took an assembly-line approach, running dozens of small projects in parallel, much like the Manhattan Project three decades later. Also like the Manhattan Project, he got results amazingly quickly—and on a scale most scientists had never considered possible. Within a few years of Haber's first drips, the BASF ammonia division had erected one of the largest factories in the world, near the city of Oppau. The plant contained several linear miles of pipes and wiring, and used gas liquefiers the size of bungalows. It had its own railroad hub to ship raw materials in, and a second hub for transporting its ten thousand workers. But perhaps the most amazing thing about Oppau was this: it worked, and it made ammonia every bit as quickly as Bosch had promised. Within a few years, ammonia production doubled, then doubled again. Profits grew even faster.

Despite this success, by the mid-1910s Bosch decided that even he had been thinking too small, and he pushed BASF to open a larger and more extravagant plant near the city of Leuna. More steel vats, more workers, more miles of pipes and wiring, more profit. By 1920 the completed Leuna plant stretched two miles wide and one mile across—"a machine as big as a town," one historian marveled.

Oppau and Leuna launched the modern fertilizer industry, and it has basically never slowed down since. Even today, a century later, the Haber-Bosch process still consumes a full 1 percent of the world's energy supply. Human beings crank out 175 million tons of ammonia fertilizer each year, and that fertilizer grows half the world's food. Half. In other words, one of every two people alive today, 3.6 billion of us, would disappear if not for Haber-Bosch. Put another way, half your body would disappear if you looked in a mirror: one of every two nitrogen atoms in your DNA and proteins would still be flitting around uselessly in the air if not for Haber's spiteful genius and Bosch's greedy vision.

———◆———

How I wish the story of Haber and Bosch ended there, simply and happily, with the two German chemists as saviors of humankind. But pride and ambition ultimately tainted each man's triumph.

Ammonia made Haber rich. His patents earned him several cents per pound produced, and BASF was soon churning out a Niagara of tens of thousands of tons per year. But Haber more or less ignored the nitrogen work after he'd sold the rights. Instead, he parlayed his sudden fame into an administrative post at the new Kaiser Wilhelm Institute in Berlin in 1911. As director, he could now hobnob with politicians and royalty, which tickled him no end. He even helped recruit Albert Einstein to come work in Berlin, and despite their diametrically opposed politics (Haber was conservative, Einstein liberal), the two became dear friends. Indeed, something about Einstein brought out a tender side of Haber. When Einstein's first marriage fell apart and his wife decamped with his two young boys, it was Haber who kept Einstein company all night as he wept.

Haber could commiserate with Einstein in part because his own marriage was unraveling. Haber and his wife, Clara, had first met as chemistry students when he was eighteen and she fifteen; he proposed fifteen years later, after inviting her to a chemistry conference as a ruse just to see her. Despite that romantic start, the marriage struggled, and Clara objected to uprooting their lives to move to Berlin. They also clashed over Haber's increasing jingoism. Even in his childhood he'd been pretty *über alles,* but his new job brought him into contact with the kaiser, and Haber developed one helluva crush. A colleague once caught Haber alone in his office practicing bowing, just in case the kaiser invited him to lunch one day. As he watched, Haber backed into and shattered a vase.

When World War I broke out, Haber transformed his institute into a minor military outpost. To be fair, in private Haber considered war futile and military leaders bozos, but he also recognized that winning the war would glorify Germany, so he joined the army and began shaving his head and wearing tailored uniforms to work. He also ringed his building with barbed wire and redirected his scientific work toward military ends. One project involved developing gasoline that wouldn't freeze during Russian winters. Another involved adapting his ammonia process to make explosives. Third, and worst, Haber began channeling his hard-won knowledge of gases into creating a new type of weapon.

Although gas warfare dates back several millennia—to the ancient Greeks, who smoked one another out of their city-states with sulfur fumes—gas attacks on the whole remained far less effective than, say, pouring boiling oil on someone. Even during the first months of World War I, when France and Russia attacked Germany with gases, the attacks almost always flopped: the French gases dispersed in the wind before the Germans even realized they'd been "attacked," while the Russian gases caught a chill, liquefied, and condensed harmlessly onto the mud.

Even if France and Russia had succeeded, their gases—all based on the element bromine—would have done little more harm than modern tear gas. Haber envisioned far more diabolical gases, based on the element chlorine. We all know chlorine from table salt and public pools, but it also forms a diatomic gas (Cl_2). And unlike calmly diatomic nitrogen, diatomic chlorine bites: its atoms have only a single bond holding them together, a bond they willingly shed to attack other atoms. Chlorine cost only a nickel per pound in Haber's day, and because it's heavier than oxygen and nitrogen, it sinks when released into the air. As a result, clouds of chlorine gas would plunge down into trenches rather than float away.

Initially, Captain Haber's gas warfare division employed just ten chemists, a modest outfit. But these ten included three future Nobel laureates, a frightening concentration of talent. Even Haber's old rival Nernst tried to pitch in, for the good of Germany. Haber muscled him aside, wanting all the glory for himself.

Several chemistry colleagues, however, saw neither honor nor glory in gas work—Haber's wife among them. Clara was a chemist herself (or had been, until her duties as a hausfrau derailed her career), and she felt that using chemicals as weapons betrayed the noble mission of chemistry, to help people. Other colleagues had more pragmatic objections. The Nobel laureate Emil Fischer hoped that Haber would fail "from the bottom of [his] patriotic heart," because he realized that if Haber succeeded, the French and British would retaliate with their own chlorine weapons.

Haber, however, argued that the Allies would never get the chance. Not because chlorine was so, so deadly. He knew that, despite its toxicity, chlorine would never kill even a fraction of the soldiers that steel would. The real tactical value of gas, he argued, was terror: enemy soldiers would see the green clouds of chlorine charging toward them and panic, fleeing the trenches and opening up the front for a German assault. One timely gas attack, he boasted, could win the whole war. Heck, he was probably *saving* lives in the long run.

In pursuit of this noble goal, Haber was willing to endure several casualties in the short term. In December 1914, one member of the gas warfare team lost his hand and another died when a test tube exploded. Haber bawled himself silly at the funeral but refused to suspend the work. Later, during field testing, Haber himself almost choked to death when he rode his horse into a contaminated area. Despite this firsthand experience with gas — or perhaps because of it — he couldn't wait to unleash chlorine on the enemy, and he finally got his chance at Ypres (in modern Belgium) in the spring of 1915, during the callously named Operation Disinfection.

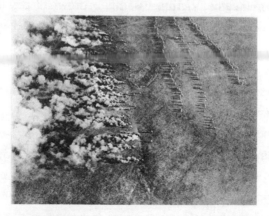

Aerial view of the first successful gas attack during World War I, near Ypres, April 22, 1915.

Like Bosch with ammonia, Haber planned his gas assault on a scale no one had dared consider before—5,730 canisters holding 168 tons of chlorine. Each canister was buried below ground level, with a hose snaking up to the surface. Contrary winds delayed the release for weeks, but on April 22—just as the German generals were losing patience—the gusts finally shifted in Germany's favor. At 5 p.m., specially trained troops crept forward and opened the valves. *Hissss.* A green storm cloud fifty feet high and four miles long began rumbling toward the French and Canadian lines.

When chlorine pierces your nostrils, it induces a reflex that makes you hold your breath. Eventually you gasp, and the invading chlorine atoms react with the water in your mouth, throat, and lungs to make hydrochloric acid (HCl) and hypochlorous acid (HClO). These acids rake the capillaries in your lungs and strip away the lining of the alveoli, the tiny sacs that absorb oxygen. There's plenty of tissue to attack here—if completely unfurled, your lungs would have enough surface area to cover a tennis court—and fluid from the broken capillaries and alveoli begins collecting in pools. This liquid blocks the flow of oxygen into your lungs, and with each second it gets harder and harder to breathe. Victims essentially drown on dry land.

The French and Canadian troops that day reacted with awe at first—they'd never seen green fog before. After one whiff, wonder gave way to terror. Horses bolted for the rear, frothing at the mouth. Foot soldiers quickly followed, hacking and stumbling, discarding rifles and even clothing as they fled. Soon the whole line broke and, just as Haber had predicted, the trenches cleared out. Unfortunately for Germany, Haber's fears about the stupidity of the German high command proved equally prophetic. Not believing the attack would work, German generals had no troops ready to exploit the breakthrough, and they brought up zero units in support. Although casualty estimates varied widely, several thousand men probably drowned that day—and all for naught, as Germany barely won any ground.

The German generals nevertheless liked what they saw and ordered a second gas attack a few days later. This too won them little, but they promoted Haber anyway and arranged for him to carry out a similar attack on the eastern front against Russia.

On the transit between France and Russia, Captain Haber stopped at home for a few days. He soon wished he hadn't. Clara confronted him, demanding that he renounce the gas work. Haber refused. To the contrary, he intended to throw a party to celebrate it, just as he had after selling the ammonia recipe to BASF.

Clara held it together during the party, but after the guests dispersed, she found herself alone. Haber had gobbled some sleeping pills and sunk down into a boozy slumber, and in the sudden quiet, Clara made a decision. She first wrote a few letters and got her worldly affairs in order. She then pinched her husband's army pistol and snuck into their garden. She fired one test shot before turning the gun toward her chest. Their thirteen-year-old son, Hermann, raced downstairs and just had time to say goodbye before she died. (Years later Hermann would commit suicide as well.)

The next morning that sentimental part of Haber broke through, and he wept over Clara's body, declaring himself shattered. However shabby a husband, he'd never quite gotten over the girl he'd loved at eighteen. But the patriot in him choked down these emotions, and he left for Russia even before her funeral. (In 1917 Haber remarried a much younger woman and started a new family. He wore a pith helmet during the ceremony.)

As predicted, the French and British soon deployed their own gas in retaliation. And in retaliation for the retaliation, Haber's team—which had mushroomed to fifteen hundred scientists by war's end—developed even nastier agents, such as mustard gas and phosgene. (If chlorine smells like swimming pools, mustard gas smells like horseradish and phosgene like cut hay.) To counter these new threats, gas masks soon became standard issue on both sides; in a pinch, soldiers could also urinate on handkerchiefs and hold them over their mouths, since the chemicals in urine neutralized the gas agents. Yet however effective at preventing lung damage, masks couldn't neutralize the terror of gas. For soldiers in the trenches, any unknown odor wafting in, even a floral bouquet in springtime, might portend a new and worse way to die. Far from ending the war, then, Haber's gases made the stalemate worse.*

——<o>——

Despite its scientific advantages Germany eventually exhausted itself and collapsed in 1918. (Ironically, one problem was a lack of food and fertilizer, since the military had commandeered most ammonia for making explosives.) Even when the bloodshed stopped, though, the hostilities continued. France had suffered more than any other country during the war, and during the peace negotiations at Versailles, French leaders were determined to make Germany howl. They demanded the equivalent of fifty thousand tons of gold as reparations, an amount equal to two-thirds of the world's gold stores at the time. (For comparison, Fort Knox holds roughly five thousand tons.)

The scientific community had its own scores to settle, and scientists in several countries denounced Haber as a war criminal for his gas research. Haber thought this absurd, and he had a point. Gas attacks caused 1.25 million casualties during the war, but just 91,000 deaths—barely 1 percent of the 8.5 million soldiers who died on the battlefield. So why condemn him and not those who manufactured shells and guns? Nevertheless, as the clamor against him grew louder, Haber fled Germany for Switzerland, where he paid a ransom for citizenship and (somewhat ridiculously) grew a beard to disguise himself. No country ever filed charges, but the taint of war crimes, like a whiff of chlorine, followed Haber the rest of his life.

Even today historians don't quite know what to make of Haber. His ammonia recipe saved millions from starvation during his lifetime, and it nourishes billions today. That makes him a genuine scientific hero, right up there with people who developed vaccines. He also invented some of the most chilling weapons in history—and unleashed them on his fellow human beings with glee. You can get moral whiplash going back and forth. Things got even messier in 1919 when Haber won the Nobel Prize in Chemistry. This cemented his status as a scientific genius, but due to public protests, he was not allowed to receive his medal at the official ceremony. He had to wait six months, and the king of Sweden did not present it to him personally.

Such snubs didn't trouble Haber. He'd done it all for his homeland, and that justified everything. In fact, he continued to devote himself to Germany after the war. To reduce the burden of reparations, he tried extracting gold from seawater, setting up secret labs in cruise ships to collect samples. The whole idea seemed nutty, but then again so had pulling fertilizer from the air.

He also continued to work on gas warfare on the sly, advising Spain and the Soviet Union. He masked this work by claiming to be studying ways to destroy existing stockpiles or convert them into insecticides. But if anyone doubted his commitment to chemical warfare—or his lack of remorse—they need only have visited his study, where he kept a framed newspaper story, reportedly signed by the kaiser, about the world's first successful gas attack at Ypres.

———◄○►———

Carl Bosch had an equally rough go of things after World War I. Following Germany's surrender in 1918, French officials demanded the right to "inspect" BASF's *ammoniawerk*—ostensibly to make sure Germany had stopped producing explosives, but really to steal

Bosch's technology. Whenever inspectors knocked on the front door, though, Bosch had his workers drop their tools and cycle all the machines down so the French could never see them in action. Bosch's workers also had the darnedest luck during inspections. Key valves and gauges kept breaking right before the inspectors arrived, and the ladders necessary to reach some machinery unaccountably went missing. One time a whole flight of stairs disappeared. Bosch never denied the inspectors access, but he rendered every visit a farce.

Thwarted and angry, the inspectors summoned Bosch—who had risen to executive status at BASF by then—to the peace talks at Versailles, where ammonia technology was a major point of contention. Upon arrival, Bosch and several colleagues were locked up in a hotel surrounded with barbed wire. After a few days of "negotiation," Bosch realized that one way or another he would have to give up the ammonia technology. Wanting the best terms possible, he risked arrest one night by scaling the barbed wire and attending a clandestine meeting with French manufacturers. There he signed a pact to help build a plant in France—provided the French left BASF alone otherwise.

They didn't. By the early 1920s reparation payments had crippled the German economy: at one point a loaf of bread cost a billion (yes, with a *b*) deutsch marks. Eventually the German government stopped paying and begged for more time. Most nations agreed, but France decided to occupy the BASF plants in Germany in lieu of payment. When Bosch again instructed his workers to shut down the machines, the French government indicted him and sentenced him, in absentia, to eight years in jail. Of the two men behind the Haber-Bosch process, then, only Bosch ended up a convicted war criminal.

Bosch regained some international respect in 1931, when he too won the Nobel Prize in Chemistry. The award surprised many, since Haber had already won for ammonia work, but the Nobel Committee rewarded Bosch more for his achievements in high-pressure chemistry than for anything else. It was fitting that the committee did so, because by that point Bosch had more or less abandoned ammonia research. BASF (now owned by a controlling company called I.G. Farben) continued to make ammonia, but with companies in other countries licensing, or stealing, the technology, profits had dwindled. Bosch had therefore channeled his knowledge of high-pressure chemistry into a new field, oil and gasoline production.

At the time, the mid-1920s, the world seemed to be running out of crude oil. (Har.) All the early gushers had run dry, and companies like BASF/Farben were scrambling to develop synthetic alternatives. Bosch grew convinced that liquefying coal and refining it into gasoline was the best option; however, the work proved tougher than expected. Scientists classify both gases and liquids as *fluids* because both gases and liquids flow under pressure. But liquids provide far more resistance to flow than gases: compare waving your hand underwater to waving it in the air. Liquid coal proved especially tarry, gumming up every pipe and valve Bosch used.

Worse, wildcatters in Oklahoma and Texas made some of the biggest oil strikes in history in the 1920s. And as oil prices began to plummet over the next decade, Bosch realized that he would never be able to screw the costs of liquid coal down enough to compete. Unfortunately, Bosch—by then director of Farben—had already staked the company's future on synthetic gasoline. Facing ruin, he began scrambling to secure government contracts to bring in money however he could. Which led him to Adolf Hitler.

Truth be told, the Nazis horrified Bosch. He especially resented the Law for the Restoration of the Professional Civil Service, which in April 1933 purged the government of Jews. As a private company Farben could and did retain its Jewish employees, and Bosch helped many other Jews find jobs overseas. At the same time, he met privately with Hitler and listened with rapt attention as the Führer waxed on and on about automobiles and his desire to secure a good

Aryan source of synthetic gasoline. (Hitler also wanted fuel for his rapidly expanding military, of course, but he was savvy enough not to tell Bosch that.) At the meeting Bosch did ask Hitler to spare some Jewish scientists, but this appeal to the dictator's decency went about as well as you'd expect. "If Jews are so important to physics and chemistry," Hitler thundered, "then we'll just have to live without physics and chemistry for a hundred years." All apologies, Bosch backed down, and to make amends afterward he praised the Nazi regime in print and attended rallies with swastika bunting and *Sieg Heil* salutes. Farben got its fuel contracts. The world got war.

By the end of World War II, factories designed by Bosch were supplying one-quarter of the gasoline in Germany. Hitler considered them so critical to the Reich that he ringed them with the best missile-defense system in Europe; even Berlin was less heavily guarded. In a real sense Bosch made the Nazi blitzkrieg possible, both by developing the technology to liquefy coal and especially by making sure the Nazis got the first drink at the trough.

To be fair, Bosch behaved no worse than colleagues in other industries, who supplied the Nazis with guns, engines, and rubber. That said, he comes off pretty rotten in comparison with his scientific peers. Einstein, Max Planck, Nernst, and other Nobel laureates — most of whom had supported Germany without reservation during World War I — refused to kowtow to Hitler. Even Haber, who had longed to see Germany rise again, reviled National Socialists as scum. In fact, just when Haber seemed to have clinched his reputation as a moral reprobate, his willingness to stand up to Hitler partially redeemed him.

As head of a government institute, Haber had to dismiss all his Jewish employees in April 1933, a crushing loss: Jews comprised just 1 percent of the German population but 20 percent of scientists. Despite his Jewish heritage, Haber himself kept his job (the law exempted World War I veterans), and for weeks he rationalized his decision to stay on, telling himself that the Nazi lunacy would pass. But on April 30 he stopped lying to himself and penned a stirring letter of resignation. "For more than forty years I have selected my collaborators on the basis of their intelligence and their character," he wrote, "and not on the basis of their grandmothers." His resignation made headlines across Germany and embarrassed Hitler no small amount: unlike, say, Einstein, no one could dismiss Haber as a mushy-headed pacifist.

The headlines did little to console Haber personally, however. After pouring so much of himself into Germany, Germany had spit him out. Heartbroken, he arranged to emigrate to Switzerland and start over, but as a final knife in the back, the Nazis confiscated the fortune he'd made from the ammonia work. Destitute, he began looking for jobs in other countries — and received zero offers. Despite his Nobel, despite his standing up to Hitler, no one could get past the stench of gas warfare on him. He finally landed an unpaid post at Cambridge University, but when he arrived there, Ernest Rutherford, the grand old man of English science, refused to shake his hand.

Facing ruin, Haber appealed to one last colleague, Carl Bosch. Although the two weren't friends, Bosch had often expressed his gratitude to Haber and had once pledged to help him if he ever needed it. "I took your words seriously," Haber wrote Bosch in 1933. "Won't you make it possible for me to live out these remaining years...in peace and decency?" Bosch never replied. With no other options Haber tried to relocate again, from England to Palestine, but his heart gave out in transit, in January 1934. His last, sentimental wish was to be buried next to Clara.

Word of Haber's pathetic end* spread throughout Europe, and when the one-year anniversary of his death came around, his old friend Max Planck decided to hold a public ceremony to honor him. Nazi officials warned Planck to cancel it, but he rented a five-hundred-seat auditorium for the memorial anyway. On the evening of the event, Planck despaired at the paucity of the crowd—mostly wives of other scientists too timid to attend. Acres seemed to separate each guest. Finally, the crowd picked up a bit as some old war colleagues of Haber's marched in. And just like with ammonia production decades earlier, the trickle turned into a torrent when Carl Bosch arrived. No doubt feeling guilty, Bosch had rallied scores of BASF employees to attend. They filled every last seat, and the stragglers had to stand in back.

Coincidentally or not, Bosch started speaking out against the Nazis, however guardedly, after the Haber rally. Unfortunately he also started hitting the bottle pretty hard and perhaps abusing painkillers. The nadir came when he gave a drunken, slurring speech defending freedom of thought at a museum opening and succeeded only in embarrassing himself. Farben finally eased him out to pasture in the later 1930s and continued to ramp up synthetic gasoline production for Hitler. Bosch died in 1940 with Germany still ascendant, but he had no illusions about how badly things would turn out. "It's a terrible gift when one can foresee the future," he said near the end. "My entire life's work will be destroyed."

Still, the future was kinder to Bosch than he imagined, though not in the short term. His magnificent miles of factories, already churning day and night for the Wehrmacht when he died, suffered heavy damage during World War II when they became major targets for Allied bombers. And the company he built became a pariah after the war for its collaboration with the Nazis. But in a more important sense, Bosch's life's work was never destroyed. It continues today, in fact thrives, and probably will until the end of civilization. Whatever their personal failings, Bosch and Haber figured out how to transform our very air into food, and it's no exaggeration to say that human beings have never invented a more important chemical reaction.

Interlude: Welding a Dangerous Weapon

Methane (CH$_4$)—currently two parts per million of air; you inhale 25 quadrillion molecules every time you breathe

The next important gas to appear in the history of our atmosphere was oxygen (O$_2$), which first began accumulating about two billion years ago. Before we get to that, however, it helps to know why oxygen is such an important gas. In short, oxygen is an instigator: it kick-starts many different kinds of chemical reactions that would never take place otherwise. One such reaction is burning. In purified form, oxygen can even start metals burning—a fact that gave a criminal in Germany an idea for an audacious heist in the late 1800s.

He was equal parts con man, cat burglar, and scientific entrepreneur. A few days before Christmas in 1890 a man calling himself

Smith checked into a hotel above the Lower Saxony Bank in Hannover, Germany. He sweet-talked the clerk into giving him one particular room — perhaps he requested a certain view — as well as the rooms on either side. (For his sister and father, he claimed, who were coming shortly.) He then retired for the evening, no doubt careful to quiet any clanking in his baggage as he climbed the stairs.

After biding his time for a few days, very early on Christmas morning, Smith pulled an umbrella and saw out of his luggage and cut a small hole in his floor. The coast was clear in the bank below, so he slipped the umbrella into the hole to catch debris, and knocked out a bigger hole. He then fished inside his suitcase for a rope ladder and lowered himself down, a sack slung over his shoulder.

Next came the cat burgling. Rather than hire a guard to protect the seven million deutsch marks in its vault, the bank had rigged up a fancy electronic security system. The vault lay in the basement and was accessible only via a spiral staircase, so bank officials wired the stairs to an alarm: the tiniest bit of pressure, even a footstep, would start the klaxons blaring. Smith had cased the joint, however, and knew he simply needed to unscrew a few electrical leads connected to the steps. The alarm neutered, he waltzed downstairs.

Now came the science. In the basement Smith pulled a few clanging canisters of oxygen out of his sack. He also pulled out a long, skinny tool consisting of two metal cylinders about eighteen inches long and a half inch in diameter; each had a rubber tube dangling off one end. He attached one tube to the bank's gaslight main, which pumped out a mixture rich in methane (CH_4). The other tube he affixed to an oxygen canister. He started the gas hissing through each, then grabbed the final item from his sack, a box of matches. With the nozzle of his cutting torch now lit, he turned to the iron vault.

Several chemists had (inadvertently) laid the foundation for this crime during the previous century, by explaining how and why substances burn. Perhaps most important, they discovered that burning requires three things: fuel, energy, and what's called an oxidizer. Oxidizers steal electrons from other substances, which is important because electrons drive all chemical reactions—chemistry is basically the study of how atoms swipe, swap, and share electrons. As its name suggests, oxygen makes an excellent oxidizer, and in stealing electrons from the methane fuel, it renders the methane unstable. The unstable methane and oxygen then react with a flash-bang-boom, undergoing a series of quick chemical changes that ultimately produce compounds called oxides (like carbon dioxide). The one caveat here is that oxygen won't start attacking the fuel without an initial kick of heat energy—hence the match. But once oxygen does start attacking, the reactions that take place will liberate more heat, which makes the process self-sustaining.

So, that's burning in a nutshell: energize some oxygen, let it attack some fuel, and stand back. But Smith needed to take a further step here, because even though he had the methane flame burning now, he still needed to get through the iron vault.

To understand this step, we can turn to the famous French chemist Antoine-Laurent Lavoisier, who discovered a funny property of iron in 1776. All metals melt at some given temperature. They also burn at another given temperature. ("Burning" here means the same thing as before, with the metal acting as the fuel this time.) In most metals, the melting temp is lower than the burning temp, but iron behaves the opposite way. It burns at around 1,800°F but melts at 2,800°F. And there's an unexpected bonus lurking in this novelty. Again, burning releases heat. So imagine making a flame hot enough to reach 1,800°F, and starting a small patch of iron burning. The heat released will warm the surrounding iron past 2,800°F and melt it. As a result, you get the extra thousand degrees for "free" in a mini–chain reaction: a small amount of burning produces a whole lot of melting.

A balding, white-bearded engineer named Thomas Fletcher finally figured out a practical application for this reaction in the later 1880s — a torch cutter that used oxygen and methane-rich natural gas. The result wasn't exactly a hot sword through butter — the cutting was sluggish — but Fletcher could nevertheless slice through iron without a blade now, a huge step forward. Indeed, Fletcher expected to make a fortune off his invention. But when he demonstrated it at a trade show in 1888, a group of safe and vault manufacturers surrounded his booth. "This method can only be used for criminal purposes," one fumed, "and should therefore be banned." They demanded that Fletcher cease and desist. He refused, and within a few years the mysterious but enterprising Smith somehow got his mitts on a Fletcher cutter.

The cutting that Christmas night took longer than Smith expected, in part because of the geometry of the hole. He was cutting a 12- by 20-inch rectangle in the vault wall. A circle of the same area would have had a smaller perimeter and therefore used less fuel. On the other hand a rectangle had a wider diagonal, allowing him to snake inside more easily if it came to that. He gambled on the rectangle, and counted on three canisters of oxygen being enough. By 1:30 a.m. he'd used up two and had nearly depleted the third.

At last, though, he worked the rectangle of metal loose. He set it aside and reached into the void, fingertips primed for soft piles of deutsch marks. Instead his hand hit something cold and hard — more iron. A double-walled vault. *Scheisse.*

Smith hoofed it upstairs and shot up the rope ladder to his room. He then gathered himself, transformed back into character, and descended to the front desk, where he explained to the clerk that urgent business had come up — just now, at 2 a.m., on Christmas — and that he had to take a train to Cologne. Smith assured the clerk he'd be back soon to grab his luggage and settle his bill. Ta-ta...

No one ever saw Smith again, but there was an interesting coda to this case. Gas-cutting technology advanced rapidly over the next few decades as engineers devised new torches and chemists discovered even more explosive gases to use. A few scientists also took a fresh look at Lavoisier's old reaction and developed a clever new technique to slice iron quickly.

This technique, oxygen cutting, involved streaming high-pressure oxygen onto a hot iron surface. As mentioned before, hot iron and oxygen will burn above a certain temperature; that is, they will react chemically and release heat and light. But there's another way to think about this process. When iron and oxygen react, they form various compounds called iron oxides. Another name for certain iron oxides is rust. On some level, then, rusting and burning are kindred operations, roughly similar chemically.*

The big difference, of course, is speed: rust can take years to skeletonize a car. But at temperatures above 1,800°F iron oxides form quickly. More to the point, iron oxides form far more quickly than iron melts. So if you want to cut a piece of steel in two by making a slice, you're better off chemically "rusting" the iron along that slice than physically melting it. And that's what oxygen cutting does—it speed-rusts along the slice. The process differs from Fletcher's torch cutting because the primary point of oxygen with Fletcher was to help ignite the methane flame; the methane flame then burns a small patch of iron, which in turn melts the iron around the patch. With oxygen cutting, you still light a flame and heat the metal. But instead of waiting around for that heat to physically melt the iron, the oxygen-cutting torch directs a separate stream of oxygen gas right onto the metal surface. This extra oxygen then kick-starts the fast chemical "rust cutting." In some sense, then, the gas itself acts as the blade here.

Admittedly, though, that's a mighty fine distinction between melt cutting and rust cutting, and quite a few captains of industry in the early 1900s happily disregarded it in pursuit of profit. You see, the oxygen-cutting technology appeared right as the world's appetite for skyscrapers and oil tankers was getting insatiable. Whoever owned the rights to the various cutting technologies could charge a bundle, and by the 1910s disputes had erupted in several countries.

One dispute involved Fletcher's old torch cutters. One side argued that Fletcher cutters simply melted the iron, end of story, so the newer oxygen-cutting patent was valid. The other side maintained that, chemical niceties aside, Fletcher's process must have involved both melt cutting *and* rust cutting; you couldn't separate the two. And if Fletcher had invented this process back in the 1880s, the patent didn't apply anymore. Unfortunately, Fletcher had died in 1903 and couldn't clear the matter up.

In the middle of the dispute, someone remembered the near-heist in Hannover. And because the plate that Smith liberated from the vault wall had ended up in a museum—it was the first attempted bank robbery with a blowtorch—a court subpoenaed the piece to examine it. You can imagine a dozen white-wigged jurists peering at the edges with magnifying glasses, scouring for flakes of rust or beads of melting. The court finally ruled that Fletcher had only melted the iron, so the rust-cutting patent stood. One of the most daring and scientific-minded crimes of the nineteenth century, then, ended up turning state's evidence and setting new legal precedent in the twentieth.

In a way, though, that outcome was fitting. Smith's main partner in this crime, the oxygen, has been setting new chemical precedents for billions of years: no other substance has so expanded the range of reactions that can take place on Earth, both in the atmosphere and especially in the bodies of living creatures. And now that we understand the power of O_2 to instigate reactions, it's time to explore where this chemical came from and just how thoroughly it revolutionized our planet.

CHAPTER THREE

The Curse and Blessing of Oxygen

Oxygen (O_2) — 21 percent of air (210,000 parts per million); you inhale roughly two sextillion molecules every time you breathe

After burning down his church in 1791, the Birmingham mob marched to Joseph Priestley's home with a spit, intending to roast him alive. You can imagine their disappointment upon finding out that he'd fled. But, lemons to lemonade, the marauders had a grand old time smashing up his furniture and destroying his chemical "elaboratory" instead. For good measure they burned an effigy of Priestley on his lawn, complete with a white wig, then beheaded it. Priestley, meanwhile, watched himself burn from a refuge on a nearby hill — one of the few people alive who, as a codiscoverer of oxygen, could potentially have explained why the fires blazed so bright.

A few years later, during the French Revolution, Priestley's biggest rival found himself a victim of mob justice as well. A decade before, Antoine-Laurent Lavoisier had articulated the connections between fire, oxygen, and breathing, declaring that breathing was a sort of slow, controlled burning in our lungs. It remains one of the most important chemical discoveries ever. But Lavoisier was also a through-and-through aristocrat: he collected taxes for the king of France and once paid the equivalent of $280,000 for a portrait of himself, his wife, and his chemistry equipment. And despite all his insights into oxygen, he lacked equivalent insight into people. In particular, he never grasped how hot the fires of rebellion can burn inside the breasts of the oppressed, and for that he faced the guillotine.

Oxygen and nitrogen are neighbors on the periodic table, and both elements form diatomic gases (N_2 and O_2) in the air. But if the buildup of nitrogen a few billion years ago gave our planet its third and most benevolent atmosphere, the arrival of oxygen inaugurated a fourth and much more explosive regime. Whereas nitrogen is non-reactive to the point of being comatose, oxygen is volatile, manic, a madman in most every chemical reaction. It actually poisons many forms of life, and caused the greatest crisis that life on Earth ever faced, the so-called Oxygen Catastrophe of two billion years ago.

Somehow, though, life turned that danger around, and this former poison became essential. It may sound trite, but this reversal always reminds me of that old chestnut about how the word "crisis" in Chinese is formed from two characters, one meaning danger and the other opportunity. Turns out that's hooey, sinologists say, but the point still stands: oxygen destroyed early life because it detonates so easily inside cells; yet when life learned how to control oxygen, that reactivity became its greatest asset. And given how much havoc oxygen has wreaked throughout history, it's fitting that this element destroyed every chemist who had a hand in its discovery. It's the Hope Diamond of the periodic table.

—————◄○►—————

The discovery of oxygen is intertwined with the most important discovery human beings ever made about air—that air consists of several different gases blended together. Before that, scientists didn't bother distinguishing between different types of gases: any fume or vapor was just "an air" to them.

A physician and alchemist named Jan Baptista van Helmont eventually set things right in the early 1600s. He was in a good position to do so. He and other alchemists had already rejected the ancient Greek system that defined air, earth, fire, and water as the four primordial elements (i.e., substances that could not be broken down further). In particular, alchemists declared that earth and fire were spurious elements—earth being a hodgepodge of materials, fire being more an event than a material thing. A series of experiments left van Helmont none so sure about air, either. He noticed that heating different substances—wood, coal, minerals—caused fumes to emerge that had different properties than air; some wouldn't burn, for instance. Similarly, he noticed distinct vapors wafting out of mines, damp cellars, and people's bellies when they belched. Van Helmont finally coined the word "gas" to deal with this miscellany, a word he adapted from the Greek word "chaos."

That etymology suits gases, considering how unruly gas molecules are, but van Helmont took things a little too far and began to speak of gases as unconquerable wild spirits. He even equated gases with souls, and declared that scientists could never confine gases to any earthly vessel. (Van Helmont apparently couldn't swim, or at least never held his breath underwater.) Later scientists stripped away van Helmont's metaphysics but retained his good idea—that air and gases are distinct things, the first being a substance, the second a state of matter.

The Irish scientist Robert Boyle took the next important steps in the study of air in the mid-1600s by investigating its physical, chemical, and biological properties. For its physical properties, Boyle explored the elasticity of air—how easily he could squish it. He eventually determined that if you compress a gas's volume, its pressure will automatically rise. (The reverse also holds true: expanding a gas's volume decreases its pressure.) As for chemical traits, Boyle noticed that flames kept under bell jars died out, as if they needed nourishment from the air. And in terms of biology, Boyle inaugurated a long and less-than-glorious tradition of scientists sticking animals beneath bell jars—larks, mice, cats, snakes, cheese mites—and taking notes while they suffocated.

Boyle mostly used ox bladders to capture and study gases, but scientists who came after him discovered that bubbling gases into tubs of water or mercury and then collecting the bubbles in upside-down flasks worked much better. With these new tools and techniques, a Scottish physician named Joseph Black made rapid progress in studying air in the 1700s.

More than any other scientist in this litany, Black simply seemed like a fun guy to hang out with. Along with Adam Smith and David Hume, he belonged to Edinburgh's famous Poker Club, a group of "literary barbarians" who drank claret and sherry by the hogshead. Black also gave lively public lectures. He collected light gases in crude balloons and let them float to the ceiling; audience members shouted that there must be invisible threads. He would then collect denser gases in containers and "pour" the invisible contents over candles, snuffing them out.

Black actually discovered this candle-snuffing gas, carbon dioxide, in 1754—the first pure gas ever isolated. He also first determined that human beings exhaled carbon dioxide, a discovery he made with a sardonic experiment in 1764. He climbed into the rafters of a local church early one winter morning and placed a beaker of slaked lime up there. Slaked lime is a fluid that produces milky precipitates when exposed to carbon dioxide. After the service wrapped up ten hours later—the Scots took religion seriously—Black snuck back up there. The preacher's hot air had indeed turned the fluid white.

Two of Black's students also made important contributions to gas theory. In 1772, twenty-three-year-old Daniel Rutherford (the future uncle of Sir Walter Scott) stripped all the reactive gases out of an enclosed vessel of air by first burning things in it and then running the leftovers through slaked lime until nothing but unreactive fumes remained. Today we call this gas nitrogen. But because it killed any mice who were trapped under bell jars with it, Rutherford called it "noxious air." (Nitrogen would also go by the names "rotten air," "burnt air," and "corrupted air" over the years. Not a popular gas.)

Black's other student, Henry Cavendish, made even greater contributions. During his lifetime, Cavendish had more money in the Bank of England than any other British subject. He also had zero friends or social life, beyond attending Royal Society meetings—and even then, he fled like a deer if anyone tried talking to him. He communicated with his domestic staff via notes, and as he lay dying in 1810, he sent everyone else out of the room so he could record the process without distraction. In retrospect Cavendish almost certainly had autism or a related disorder, but in any case, England has rarely produced a finer mind. He measured the density of Earth (and therefore its mass) without ever leaving his mansion in London, using little more than four lead balls and some wire. He also discovered hydrogen gas in 1766 by letting acids sizzle on top of different metals. Cavendish called hydrogen "inflammable air" because flames roared up when exposed to it.

Beyond isolating it, Cavendish made another vital discovery about hydrogen. He mixed hydrogen with common air in a glass container one day and began sparking the mixture with a newfangled (and no doubt enormously expensive) electric genera-tor. Afterward a "dew" formed on the inside of the glass. This dew had nothing to do with the original experiment, but like great scien-tists throughout history Cavendish pursued this anomaly. To his shock, he discovered the dew was water. Cavendish couldn't have understood all the details here, since modern chemical theory didn't exist then. But he did grasp that two gases — and two flammable gases at that, air and hydrogen — had combined to form liquid water, pretty much the opposite of flames.

Even today that's hard to wrap your head around; back then it boggled the mind. And in a broader sense, this experiment killed off the final classical element of ancient Greece. Earth and fire had been exposed as pretenders several centuries earlier. Air held on until experiments on gases began in earnest in the mid-1600s. Water had so far survived these purges, but a few drops of dew and a keen experimental eye had unmasked it as a composite. Cavendish had finally toppled a system that had dominated natural science for two thousand years. But now something had to replace that system. And with so much prestige at stake, the competition among scientists got hot enough to burn everyone involved.

◄○►

Although oxygen is the most important element in the air around us, no one knew it existed until an apothecary in Sweden named Carl Scheele began roasting various minerals in the early 1770s and col-lecting the fumes they emitted. Because candles burned so brightly in these fumes — now known to be oxygen — he called the gas "fire-air." Scheele also discovered that animals breathing this gas inside bell jars lived *longer,* a refreshing change from the carnage of every other gas.

For reasons that remain unclear, Scheele sat on his results for years, then picked a lazy, crooked printer who delayed publication for several more. His work finally appeared in 1777 — years after two other scientists had already made claims for oxygen. Unfortunately, Scheele was in no shape to defend his priority over the next decade, thanks to two bad habits: he worked in unventilated rooms, and he tasted every chemical he made, even ones like mercury cyanide. (Which is a bit like shooting yourself with a radioactive bullet — multiple ways to die.) Not surprisingly, Scheele expired, unknown and uncelebrated, at age forty-four — the first casualty of the oxygen curse.

Chemist Joseph Priestley, codiscoverer of oxygen and eponym of the Priestley Riot.

Most of the historical credit for discovering oxygen fell instead to a preacher and rabble-rouser named Joseph Priestley. Priestley was the firstborn of six children, and his mother died when he was six years old. (In case you missed the math there, that means she had six children in six years; no wonder she blew a gasket.) Afterward Priestley went to live with his aunt, who belonged to a liberal sect of Protestants, and she steered her talented nephew into the ministry.

To her eternal regret, Priestley turned radical shortly after taking the cloth, even questioning the divinity of Christ. He soon became infamous across England, and cartoonists delighted in caricaturing him. Some portrayed him as a woodpecker, mocking his beakish nose and stutter; others simply slapped horns on him and called him Satan. As a result Priestley and his family bounced from parish to parish throughout his career. He was always secure in his convictions but rarely in his finances.

Priestley began dabbling in science for a few reasons. First was curiosity — he wanted to understand God's creation. Second was camaraderie, since scientists didn't shun him for his unorthodox theology. He even befriended Benjamin Franklin, who stoked Priestley's first scientific interest in electricity. (While writing a book about electricity in 1769, Priestley decided to sketch the diagrams himself but had a devil of a time getting them right. Reluctant to throw away paper, an expensive item back then, he searched for a way to strip the pencil lead off the surface. He eventually hit upon India rubber — the first rubber eraser.)

Priestley also got into science, believe it or not, to make money. He bought some math books and globes with an eye toward tutoring students, then tried the gallant profession of science writing. Strangely enough, science writing proved an expensive habit: the books he needed for research cost a bundle. So he began experimenting and writing about his experiments as a cheaper alternative. (Fancy that, scientists of today: equipment and supplies cost *less* than books.)

Priestley discovered his scientific passion — the study of "airs" — after transferring to a parish in Leeds in 1767. He happened to move next door to a brewery, and the gas bubbling out of the beer vats (carbon dioxide) fascinated him. He eventually got himself banned from the brewery after spilling some ether into a vat during an experiment and ruining the batch. (The congregation didn't like him spending all his time there anyway.) Priestley quickly shifted his work on airs to what he called his private "elaboratory."

As a poor preacher, he had to conduct these experiments with everyday equipment.* He'd heat substances in a gun barrel, the test tube of its day—a hardy container that wouldn't crack. To capture any gases emitted, he'd fit a tobacco pipe into the muzzle and channel the vapors into his washerwoman's tub or a teakettle. As gas bubbled through the water there, he'd collect it in a bottle and turn it loose on mice trapped beneath beer glasses. If something went wrong, stinking fumes would flood his home, sending his family scattering. But science is forever grateful that his wife and children displayed such Christian forbearance: Before 1770, scientists knew of only two pure gases, carbon dioxide and hydrogen. Priestley alone discovered nine new airs over the next few decades, including ammonia, sulfur dioxide, and nitrous oxide (laughing gas).

After his stint in Leeds, Priestley found a patron—William Petty, Lord Shelburne—to support his scientific habit, and in August 1774 he conducted the experiment that eventually made him famous. It must have been a beautiful thing to see. Using a twelve-inch lens, he focused some (rare) English sunlight onto a glass jar; inside the jar sat a small pile of red powder (mercuric oxide, HgO). As the beam of sunshine heated the red powder, liquid mercury would have bubbled out, sprouting on the surface and blossoming into tiny quicksilver spheres. Meanwhile, a clear gas steamed upward from the red powder, shimmying in the air. Oxygen.

It's not always clear what Priestley understood about this gas, but he definitely didn't think he'd discovered a new element. Rather, he described oxygen in terms of a hoary scientific theory called phlogiston. Now, the phunny name aside, phlogiston is actually a fascinating episode in science history—a theory that, while wrong, nevertheless pushed science in creative directions. But rather than digress for thirty pages, I'll mention just two salient phacts for this story. One, scientists thought of phlogiston as an invisible but *material* substance that was released into the air whenever things burned. Two, Priestley believed in phlogiston even more fervently than most scientists, and he interpreted all his discoveries in terms of it. This included oxygen, which he dubbed "dephlogisticated air." (Don't ask.)

Like Scheele, Priestley found that candles burned brilliantly in dephlogisticated air. He also rejoiced to see that it didn't smother mice. In fact, the mice were having such a grand time that Priestley decided to take a few hits of the gas himself, and it proved every bit as lovely as he'd hoped. "I fancied that my breast felt peculiarly light and easy," he reported. "In time, this pure air may become a fashionable article in luxury." (And to think: Michael Jackson was sleeping in a hyperbaric chamber a mere two centuries later.) Rhapsodies aside, these experiments did reveal a deeper link between burning and breathing, since this new gas encouraged both processes.

Unlike Scheele, Priestley never lost any time in publicizing his work. So when Lord Shelburne offered to pay his passage to Paris in October 1774, just a few months after the first oxygen experiments, Priestley arranged to dine with several savants from the French Academy of Sciences and spill everything he knew. In a life full of mistakes, it was the biggest he ever made.

———◇———

History doesn't record the menu, but given the milieu—the decadence of prerevolutionary Paris, the mansion of one of the richest couples in France—we can imagine the spread: roast duck with

Chemist Antoine-Laurent Lavoisier, who revolutionized his field and was executed during the French Revolution.

truffles, ham in champagne sauce, ragout of carp tongues, quinces and bonbons and creams. Still, for Antoine-Laurent Lavoisier and his wife, Anne-Marie, the cuisine took second place that night. They listened with rapt attention as their guest, Joseph Priestley, blathered on about a potent new air he'd liberated from a red powder.

Slyly, Lavoisier declined to mention that he'd performed a similar experiment one month prior. He hadn't noticed any gas escaping, but as soon as Priestley left Paris, that gas would become his obsession—to the point that he would try to steal credit for discovering it. They'd met in friendship over dinner, but oxygen would soon make enemies of Lavoisier and Priestley.

Lavoisier had one huge advantage in this rivalry: money. He'd been born rich; had become even richer when he'd married into Anne-Marie's family; and had become richer still when he bought a half share in the Ferme Générale, a private company that collected taxes for the Crown. Each year the king's advisors told the Ferme how much revenue the government expected from goods such as barley and tobacco. The Ferme then shook down farmers and tobacconists across France for money—and kept all excess revenue as profit. Thousands of goods fell under their purview, and Ferme members didn't even have to abuse the system: it was designed to make them obnoxiously wealthy. It was the best investment Lavoisier ever made—he cleared the equivalent of five million dollars some years—and also the worst, since it cost him his life.

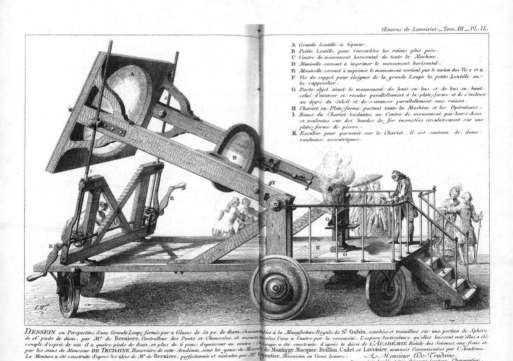

Lavoisier's gigantic lens for burning diamonds.

In 1772, the year he turned twenty-nine, Lavoisier began pouring a large fraction of his wealth into science. He became most famous for burning up diamonds with sunlight, a feat he accomplished with a gigantic lens mounted on a wheeled platform; it looked like a telescope crossed with a siege engine. Although wasteful—who but an ancien régime aristocrat would burn diamonds?—it did show how incredibly powerful sunlight is. (Think about frying diamonds the next time you're tempted to stare into the sun.) It also got him thinking about the underlying nature of matter: did the diamond disappear when it burned up, or was it simply converted into a different state?

That same year Lavoisier came across the work of Priestley, who was ten years older. Priestley had invented carbonated water a few years before by bubbling CO_2 through H_2O. The French navy thought this fizzy drink might cure scurvy, and they asked Lavoisier to investigate.* That theory didn't hold water, but Lavoisier came away impressed with Priestley. Coincidentally or not, he himself began working on gases shortly afterward.

In doing this work he collaborated with his wife. Anne-Marie had lost her mother at a young age and had been raised in a convent, one of the few places where a girl could get an education back then. She left to marry Lavoisier, and despite the difference in their ages (she was thirteen, he twenty-eight), Lavoisier treated her as an intellectual equal to a large degree. In particular, because she spoke several languages, he asked her to translate some papers on phlogiston into French. In the process she highlighted inconsistencies in the theory, and Lavoisier agreed that phlogiston smelled fishy.

In November 1772 he hit upon hard proof of this. As a material substance, phlogiston should have had weight. And when things like firewood and candles burned—and supposedly lost phlogiston— they did indeed lose weight. But when Lavoisier burned metals with the huge lens, they actually *gained* weight, which made no sense. Some chemists explained this anomaly by suggesting that phlogiston sometimes had "negative weight," but most scientists realized how stupid that sounded. Still, the discrepancy couldn't kill phlogiston theory off, because no one had a better explanation for why things burned—until Lavoisier heard Priestley babbling over dinner.

Lavoisier suspected that Priestley had liberated a new gas without realizing it, and he retreated to his laboratory to investigate. (Oddly, this lab—located in the Paris Arsenal—had a gallery for spectators; Lavoisier actually had an audience for experiments at times.) Lavoisier isolated the gas, which he eventually named oxygen. And after a few experiments, he determined that oxygen explained the chemistry of burning much more neatly than phlogiston could. To wit, burning always seemed to involve oxygen combining with a substance and releasing heat and light. Carbon in wood and coal, for instance, combined with oxygen to form carbon dioxide, which then floated away. Oxygen combined in a similar way with the atoms in burning metals. But unlike with carbon dioxide, the metal–oxygen compounds were too heavy to float away. This explained why metals gained weight when they burned—they absorbed oxygen. This insight rendered phlogiston superfluous, since Lavoisier could explain burning without it.

Lavoisier's next round of experiments, on oxygen and breathing, must have been far more entertaining for his audience. Less so for a colleague of his, Armand Séguin. For several days in a row Lavoisier mummified Séguin in taffeta and painted him with rubber latex to seal him in. He then glued a tube to Séguin's lips and connected it to a bottle of oxygen. The actual experiment consisted of Séguin doing different tasks—sitting still, digesting a heavy meal, pumping a foot treadle for exercise—while Lavoisier monitored how much oxygen he consumed and how much carbon dioxide he exhaled. The work revealed a clear correlation between gas consumption and breathing: the harder Séguin worked, the more oxygen he sucked

Lavoisier once mummified a colleague to study breathing. (Photo courtesy Wellcome Trust)

down and the more carbon dioxide he huffed and puffed. Again, no need to invoke phlogiston to explain breathing.

(Incidentally, these experiments provided the first real answer to an ancient conundrum: When you exercise and lose weight, where does the "lost" weight go? Most of it gets released into the air as gases like carbon dioxide—further proof that you came from gas, and unto gas you shall return.)

Marshaling all this evidence, Lavoisier presented his findings on oxygen at a scientific meeting in Paris in September 1775, just a year after Priestley had tipped him off. It was an Apollonian performance, suffused with facts and elegant logic, and it transformed chemistry forever. But once he'd finished, Lavoisier and his wife got a little naughty and indulged their Bacchanalian side. First, they hired actors and staged a mock trial involving characters named Oxygen and Phlogiston. (No points for guessing which was the hero and which the dunce.) Madame Lavoisier then dressed up like a pagan priestess and burned several textbooks on phlogiston, to exorcise it. When the Lavoisiers dispatched a theory, they did it in style.

Not everyone applauded. Priestley fumed that Lavoisier was hogging credit for oxygen, and the debate about its true discoverer continues today. There's no question Scheele isolated and collected it first, followed independently by Priestley. There's also no question that Lavoisier had never noticed oxygen until Priestley alerted him. To some historians the discussion ends there, with Lavoisier shut out. On the other hand, neither Scheele nor Priestley understood what he'd found. Imagine a fossil hunter unearthing a skull and announcing that he's found an ancient primate of some sort; now imagine a paleontologist catching a glimpse and realizing that it's actually something better—an extinct species of human being, a whole new branch of humanity. Speaking technically, the first guy discovered the skull, but only the second understood it. Lavoisier stands in a similar position vis-à-vis Priestley and Scheele.

And in some ways, debating who discovered oxygen misses the point. Lavoisier cared less about oxygen per se and more about what oxygen revealed about chemistry in general. Several chemists before him, for instance, had suggested that breathing and burning shared something in common, but only Lavoisier identified oxygen as the gas that supported each. Historians have also faulted Lavoisier for copying other chemists' work, but this overlooks the fact that he almost always surpassed them in doing so. As Henry Cavendish had, Lavoisier once sparked oxygen and hydrogen in a container and produced a watery dew. Also like Cavendish, Lavoisier compared the mass of the two gases before the reaction with the mass of the water afterward and found them equal. (Lavoisier had constructed a ridiculously accurate balance for the purpose, with a precision of 0.0005 grams.) But unlike Cavendish, Lavoisier saw the bigger picture here: he reasoned that the same equality should hold true in *any* chemical reaction: the mass of the products should always equal the mass of the reactants. And from there he developed a general law of chemistry, the law of conservation of mass. It says that in any reaction, even if the original substances change color or phase or recombine in strange ways, there must, must, *must* be the same amount of stuff before and after. Mass is conserved.

(Curiously, some historians have argued that Lavoisier's work collecting taxes for the Ferme Générale probably inspired his science here. Figuring out the law of conservation of mass involved weighing lots of reactants and products, and getting everything to even out on the chemical "balance sheet." Accountants track debits and credits the same way—in either case, dollars and atoms can't just disappear. So for any readers out there with horrid memories of balancing equations in chemistry class, blame the dismal science of economics.)

That's just the start of Lavoisier's contributions to chemistry. He was the first scientist to explain that one substance—say, water—can appear as a solid, liquid, or gas, depending on its temperature. He also proposed that all substances are either elements (e.g., oxygen, carbon, and iron) or combinations of elements—probably the most fundamental idea in chemistry. As one nineteenth-century chemist said, "Lavoisier discovered no new body, no new property, no natural phenomenon previously unknown. His immortal glory consists in this—he infused into the body of science a new spirit."

Lavoisier presented this new spirit in his *Traité Élémentaire de Chimie,* chemistry's answer to *Principia Mathematica* and *On the Origin of Species.* By a wonderful coincidence, *Traité* debuted in 1789, the year of another revolution in France, and historians have found it irresistible to point out that Lavoisier's chemical revolution would, in time, shake the world just as profoundly. Unfortunately, Lavoisier himself would not live long enough to see the fruits of either.

———<◦>———

Most of the oxygen in the air right now comes from photosynthesizing plants and bacteria—scientists understand that perfectly well. What they don't fully grasp is how organisms started producing oxygen in the first place, and how that process transformed the nitrogen-heavy atmosphere of yesteryear into the air we breathe today.

The first living organisms probably arose near underwater volcanic vents and probably used the sulfur there to drive their metabolisms. We call such creatures anaerobic bacteria,* and similar microbes still exist today. (They contribute to morning breath, for one thing.) About three billion years ago, however, one line of anaerobic bacteria evolved into cyanobacteria, which power themselves not by underwater volcanic heat but sunlight. Cyanobacteria (a.k.a. blue-green algae) weren't the first microbes to harvest light for energy, a process called photosynthesis. In fact, several sun-dappled species of microbes likely existed then, each one colored differently to collect different wavelengths of light. (I imagine the beaches and tidal flats of the time looking like Rothko paintings, with intermingled mats of purples, greens, and reds.) No, what made cyanobacteria unique wasn't using sunlight per se but using sunlight to strip electrons off water molecules. As we saw before, electrons drive chemical reactions. And rather than rely on somewhat scarce elements like sulfur for electrons, cyanobacteria could now harvest electrons from one of the most ubiquitous molecules on Earth, vastly increasing their rate of production.

The process of converting sunlight to useful energy starts with chlorophyll, a green colored molecule that acts like a biological antenna and absorbs light from the sky. After that, the biochemistry gets a little convoluted. But cyanobacteria basically use sunlight to break certain molecules down, and then create other molecules to store energy for later use. For example, cyanobacteria might break down water into H_2 and O, then fuse some of those fragments with CO_2 and other molecules to make sugars like glucose.

There's one other thing to keep in mind with photosynthesis, however, something that Antoine-Laurent Lavoisier—obsessed as he was with chemical accounting—would have noticed right away. It takes 6 molecules of carbon dioxide and 6 molecules of water to make 1 molecule of glucose: $6CO_2 + 6H_2O \rightarrow C_6H_{12}O_6$. But notice that the sugar ($C_6H_{12}O_6$) contains only 6 oxygens; more to the point, it contains just 6 of the 18 oxygens available at the beginning. Conservation of mass says that atoms can neither be created nor destroyed, so we can conclude from this that photosynthesis must produce free oxygen gas (O_2) as a by-product. And in fact, as cyanobacteria began to thrive and spread across the globe, free oxygen began to accumulate for the first time in Earth's history.

Free oxygen seems dandy today, but organisms back then found it toxic. The problem is that when ultraviolet light strikes O_2, the gas mutates and forms free radicals, which chew through DNA and proteins and shred them. Free oxygen also destroyed the ability of many nitrogen-fixing bacteria to do their thing, since oxygen tears out the iron atoms at the heart of the nitrogenase enzyme. Oxygen was profoundly antilife back then.

Luckily for these delicate microbes, free oxygen had trouble accumulating at first, since it reacted with pretty much everything it encountered in the water and air (nitrogen gas being an exception). In particular oxygen in the ocean reacted with dissolved iron there to form rusty precipitates. These microflakes then wafted down to the ocean floor and accumulated over the millennia into so-called banded-iron beds, reddish layers that still exist in rock outcroppings worldwide, wherever an ancient seabed from this era has been lifted up to dry land. These layers still contain 90 percent of the world's iron reserves, all thanks to microbes.

While this dissolved iron lasted, oxygen pollution was minimal. But year by year, epoch by epoch, cyanobacteria stripped the ocean of iron, and once they'd depleted it, things got ugly. Free oxygen began accumulating in the seas, slowly strangling life there. It then bubbled upward and invaded the atmosphere, the microbial equivalent of the clouds of deadly gas at Lake Nyos. Scientists call this buildup of oxygen over several hundred million years the Great Oxygenation Event, and like the Big Bang that started our universe and the Big Thwack that created our moon, that name is a monument to understatement. Life never faced a graver threat—describing it as a holocaust isn't hyperbole. Every last branch and twig of life was staring hard at extinction.

I'm going to keep you in suspense for a little longer by not revealing whether life on Earth pulled through. But I can reveal that several microbes began to fight back against this gaseous brute. Some developed tougher outer membranes to keep oxygen out altogether. Some built special interior walls to shield delicate molecules like nitrogenase. And some microbes did nothing and got lucky: when oxygen rushed through their membranes, instead of dying an agonizing death, their cellular machinery realized it could exploit this gas for energy, channeling its explosive power toward productive ends. Imagine a few French troops breathing in the chlorine from Haber's gas attack and not only surviving but feeling *très* invigorated—it's about as likely. But these special bacteria, now called aerobic bacteria, somehow pulled it off.

We animals owe aerobic bacteria a debt of gratitude, since they made multicellular creatures like us possible. All animals' cells contain little critters called mitochondria that descended from these aerobic pioneers, and it's mitochondria that allow our cells to utilize oxygen. Indeed, mitochondria are the key to understanding the link between oxygen and higher life. I mean, every schoolchild knows that animals need oxygen to live, but we rarely learn *why* animals need it. The short answer is that mitochondria use oxygen to break down and extract energy from sugars like glucose. True, our cells can digest glucose a little bit without oxygen. But to really strip every last erg of energy from glucose—to suck its marrow— mitochondria have to go to work on glucose with O_2. Without it, our batteries would run down and we'd die within seconds.

(I should note, too, that despite what you might remember from grade school, plants also breathe in oxygen. The usual line is that plants breathe in carbon dioxide and emit oxygen, while animals breathe in oxygen and emit carbon dioxide. And that's true as far as it goes, since plants do expel oxygen while making sugars. But making sugars is only part of what plants do. They also grow and reproduce and fight off predators, and to do those things, they need oxygen, which they "inhale" through pores in their skin.* Plants even employ the same mitochondria we do to handle O_2 in their cells.)

Beyond mutating the tree of life, oxygen also transformed our planet's surface. Once enough free oxygen accumulated in the air, it started attacking greenhouse gases like methane, removing them from circulation. This in turn sabotaged our planet's thermostat and caused surface temperatures to plummet—so much so that we might have entered a "snowball Earth" phase, with glaciers forming even near the equator. At the same time, oxygen also beautified the planet. Something like two-thirds of the 4,500 minerals on Earth today can form only in the presence of oxygen. This includes several precious gems such as turquoise, azurite, and malachite—jewels we wouldn't have without oxygen. By that same token, several known minerals cannot form in the presence of oxygen—meaning that oxygen essentially drove those ancient minerals extinct. Rocks, then, can spread, evolve, and die out just like species can, all depending on the gases they "breathe." The legendary biologist Carl Linnaeus, who invented the two-part naming system that scientists use for plants and animals (*Homo sapiens, Tyrannosaurus rex*), actually included minerals in his original schema. Later biologists shunned minerals as irrelevant, but oxygen is potent enough to kinda sorta stir rocks to life.

Oxygen now makes up 21 percent of our air—with roughly half that amount coming from plants and the other half from microbes. But unlike nitrogen, the other dominant gas in the air, oxygen didn't accumulate steadily over the billennia. It spurted. The first spurt started around 2.3 billion years ago, after the iron in the ocean ran out. Over the next several hundred million years, the concentration of oxygen in the air jumped from one molecule per trillion to one molecule per five hundred—almost ten orders of magnitude. For some perspective there, some of Haber's gases can kill humans at parts per million, so the rise to one in five hundred made things pretty noxious for non–oxygen breathers.

Starting around 1.8 billion years ago, oxygen levels paused and held steady for a while, mostly because minerals on land absorbed it. Probably not coincidentally, life, too, seemed to stall out and quit evolving during this period. (Geologists sometimes refer to this time as the "boring billion.") But concentrations started creeping up again about 600 million years ago, when the minerals on land became saturated, and with this rise, the first complex plants and animals began to appear in the fossil record—creatures capable of running, fighting, hunting, mating, killing.

In the hundreds of millions of years since, oxygen levels have veered drunkenly, dipping as low as 15 percent and rising as high as 35 percent. Again, this had several unusual effects. During the highs, the slightest spark or cinder would have flamed up and burned everything in its vicinity. During the lows, even volcanoes and lightning strikes would have struggled to ignite anything nearby. We in fact see zero evidence of fire in the fossil record until a few hundred million years ago, when the first hints of black charcoal appear. This also marks the first point in history when a time traveler could have stepped out of her time machine and walked around without gasping. (At anything below about 17 percent O_2, our thinking gets foggy and we have trouble moving.)

Of all animals, insects probably benefitted the most from high oxygen levels. Because insects lack lungs, most of them, especially small ones, can't really inhale oxygen; it passively diffuses into their cells instead, through pores in their exoskeletons. This setup works fine — until insects grow too large. It's a geometric fact that surface area increases more slowly than volume, and at some point their little pores can't suck down enough O_2. This explains why most insects nowadays are tiny: they'd suffocate otherwise. Back in the heady days of 35 percent oxygen, however, that constraint didn't matter so much. Had our time-traveling friend popped out of her wormhole 300 million years ago, she would have confronted millipedes a yard long, dragonflies the size of seagulls, and spiders as wide as tires. They were the colossi of insects, all thanks to oxygen.

At current oxygen levels, a human being needs to take one breath every four seconds, around twenty thousand breaths each day. That means that each of us burns through one septillion (1,000,000,000, 000,000,000,000,000) oxygen molecules every twenty-four hours. With seven billion people on Earth, as well as a bazillion other organisms that need oxygen, you can see what a greedy lot we animals are. If every plant and O_2-making bacterium on Earth disappeared tomorrow, human beings and other animals would probably choke to death within a thousand years—ending intelligent life on Earth in less than a millionth of the time it took to evolve.

Thankfully, plants and cyanobacteria replenish our oxygen budget every day. Indeed, if you look at the give-and-take between different forms of life, everything balances beautifully. Again, plants generally take in carbon dioxide and water and make sugars and oxygen; animals generally take in sugars and oxygen and make carbon dioxide and water. Yin and yang, thesis and antithesis, perfect equilibrium. Not even a stickler for chemical accounting like Lavoisier could find fault. You hear lots of talk in physics about symmetry and the beauty of nature deep down. All true. But for my money the O_2/CO_2 symmetry is more inspiring, because it took so long to evolve and because there are so many more moving parts, so many more ways it could all go wrong. Yet it doesn't. Oak trees, birds of paradise, cyanobacteria, we're all in this together.

———◦———

It's hard not to pity Joseph Priestley. Lavoisier already had all the advantages of rank and station, and he proceeded to rip off Priestley's experiments and horn in on the discovery of oxygen. Worst of all, Lavoisier then had the gall to twist Priestley's own discovery against him, and use oxygen to dismantle his beloved theory of phlogiston. The real bitch of it was that Lavoisier succeeded. Each year Priestley found himself with fewer and fewer allies, while the ranks of those loyal to Lavoisier swelled.

Priestley soon had graver things to worry about than his scientific honor, however. Although known throughout England as "Dr. Phlogiston," he had always considered himself a preacher first, and in 1780 he landed a pretty good God gig in Birmingham. He later called his time there the happiest of his life. He attracted a well-to-do congregation; he discovered more new gases; he befriended James Watt and Erasmus Darwin, and joined their famous salon, the Lunar Society; he even made some real money by building and selling scientific equipment.

Not everyone in Birmingham welcomed him, though. The city had endured a series of riots throughout the 1700s, and hoi polloi there had earned a reputation as a "beggarly, brass-making, brazen-faced, brazen-hearted, blackguard, bustling, booby Birmingham mob." It didn't take them long to single out the new preacher for attention. Priestley had written several notorious books, including the bluntly titled *History of the Corruptions of Christianity*. He had also publicly cheered the French Revolution, which your typical Birmingham boobies associated (not unjustly) with death to church and king, two beloved institutions. (Nor were they alone in despising the Revolution. Statesman Edmund Burke, in groping for a smart metaphor, compared the deadly Paris mobs to a "wild gas"—invoking van Helmont's vision of gas as pure chaos.)

Thumbing their noses at such sentiment, ninety of Priestley's friends in Birmingham planned a lavish dinner at a local hotel on July 14, 1791, to celebrate the second anniversary of the storming of the Bastille. Bad idea. While they feasted and raised glasses to every radical notion they could think of—"patriots of France," "the rights of man," freedom fighters in America—three hundred thugs and lowlifes gathered outside; the town crier ran about clanging his bell. When dinner broke up, the mob threw stones at the guests and smashed the hotel windows. They then marched to Priestley's church, tore out the pews, and burned the building to cinders. Anger still unslaked, they stamped the mile out to Priestley's home to roast him alive.

The destruction of Joseph Priestley's home during the Priestley Riot.

Priestley had shown uncharacteristic prudence in not attending the dinner. He was in fact playing backgammon when he heard the approaching roar. Friends arrived to hustle him away, and he watched the destruction of his home and the burning of himself in effigy from a nearby hill, close enough to hear one rioter shout, "Let's shake some powder out of Priestley's wig!" The boobies destroyed two dozen more homes and burnt down four more churches over the next three days; much of Birmingham was reduced to charred timber and graffiti. The event nevertheless went down in history as the Priestley Riot, after its most prominent target—the only riot, so far as I can tell, ever named after a scientist.

After relocating to London, Priestley applied for recompense from the government, which had stood by and done nothing to quell the mob, but the Crown showed little sympathy. "I cannot but feel better pleased that Priestley is the sufferer for the doctrines he and his party have instilled," George III declared. Priestley meanwhile tried to reconnect with old friends at the Royal Society in London — and found himself shunned, a rejection that hurt him deeply. (So much for camaraderie in science.) The courts eventually awarded him around £2,500 in damages, £1,500 short of what he'd sought. Much of that discrepancy involved his scientific equipment, which he valued highly and the court considered worthless.

With nothing left for him in England, Priestley sailed for the United States in April 1794, at age sixty-one. All three of his sons had already emigrated there, and they assured him that the rumors of religious freedom were true. Despite a generous offer to teach chemistry at the University of Pennsylvania (founded by his old friend Benjamin Franklin), Priestley settled in rustic Northumberland in central Pennsylvania instead. After landing in America, he never again wore his powdered wig, intent on leaving the Old World and its troubles behind.

Trouble, however, stalked him across the Atlantic, when an old political enemy and fellow expat (writing as "Peter Porcupine") began distributing scurrilous pamphlets in Pennsylvania to ruin his reputation among his new neighbors. Priestley also endured a domestic scandal when his son William was accused of poisoning several servants including two young girls. (The case was never resolved.) Amid the turmoil Dr. Phlogiston continued to experiment, and he discovered one last gas, carbon monoxide, on these shores. When he finally died in 1804, President Thomas Jefferson — no scientific slouch himself — eulogized him as "one of the few precious minds to mankind."

———<o>———

Lavoisier met a grislier end. During the French Revolution a radical faction known as the Jacobins seized power in France and abolished the Ferme Générale in 1790. Suddenly jobless, Lavoisier turned his attention toward scientific work like developing the nascent metric system. But the stain of tax collecting would not wash away, and people began grumbling about revenge. An influential radical named Jean-Paul Marat then turned Lavoisier's scientific skills against him. Years earlier, to help the Ferme collect taxes, Lavoisier had recommended building a wall around Paris and installing toll gates to control the flow of traffic in and out. A modern historian compared the idea to "forty of the wealthiest individuals in the United States walling in New York City and building palatial toll gates, at taxpayer expense, for use by the IRS." The Ferme of course loved the idea, and in exacting the tolls they earned the everlasting hatred of every sans culotte: years later, even before they stormed the Bastille, the mob made a point of attacking and burning the hated gates. Marat took the damnations even further by accusing Lavoisier of building the gates to block the flow of *air* into Paris— as if Lavoisier could control the very oxygen that people breathed. As science, the idea was hooey. As propaganda it was brilliant.

However hated, Lavoisier remained a free man until 1793, when the Jacobins finally issued an arrest warrant for him and the twenty-seven other "leeches" and "vampires" of the erstwhile Ferme. Lavoisier hid out in the Louvre for a few days before turning himself in on Christmas Eve. The first prison he landed in was actually pretty cushy; he and his fellow prisoners enjoyed wine and pears and backgammon. During this phase, many Ferme members still held out hope: Lavoisier consoled himself with the notion that even if he lost his fortune, he could always become a pharmacist, his favorite profession. But the prisoners realized just how dire things were when they got transferred to more squalid cells, which exposed them to rats and fleas probably for the first time in their lives. Suspecting their fates, some Ferme members secured lethal doses of opium, but Lavoisier talked them out of suicide.

Their trial commenced at 10 a.m. on May 8, 1794. The three judges each wore black robes, white ties, and hats with feathers; they also had a bottle of wine to sip, which made it seem like a civilized affair. As he entered, each prisoner was stripped of all personal belongings—Lavoisier had a small gold key confiscated. The judges proceeded to detail, for the first time, the exact charges against them: fixing interest rates, watering down tobacco, and embezzling 130 million livres (equal to $5 billion today). Most ominously, the government charged the Ferme with conspiring against the people of France, a capital crime. None of the charges held up under Lavoisier's scrutiny—it was like the mock trial between Oxygen and Phlogiston all over again. But that kind of logic made little headway in this court, and the judges sentenced all twenty-eight defendants to the guillotine.

The condemned rode to the guillotine in carts that very day, with soldiers on horseback clearing a path through the jeering crowds. About 5 p.m. they mounted the scaffold with hands tied behind their backs, and genuflected one by one before the blade. Lavoisier went fourth. "It took only a moment to cut off that head," one friend mourned, "and a hundred years may not give us another like it." Joseph Priestley was halfway across the Atlantic just then, and would not hear the news for months. But when he did, it must have struck him that his beloved French Revolution had now destroyed not just his life but the life of his greatest scientific rival.

In the years after his death, several legends arose about Lavoisier's final hours. One held that, after receiving his death sentence, he requested a short stay of execution to wrap up some scientific work. The judge allegedly dismissed him, declaring, "The Republic has no need of scientists." This never happened—but some people do think this way. Adolf Hitler voiced much the same sentiment when Carl Bosch asked him to spare a few Jewish scientists.

The more famous legend involves Lavoisier's last moments. At some point while queuing up for the guillotine, Lavoisier supposedly asked a colleague to push as close as possible to the scaffold and help with one final experiment. Contrary to what you might expect, the guillotine doesn't kill you instantly; the head lives on for several seconds after being severed. For obvious reasons, no one had ever tested exactly how long it lives, and Lavoisier decided that his own death provided a fine opportunity. So he informed his assistant that, as soon as he felt the whisper of the blade on his neck, he would begin blinking, and would continue blinking until his internal supply of oxygen ran low and he lost consciousness. All his assistant had to do was ignore the blood and the mob and his own revulsion, and keep count. Some sources claim Lavoisier snuck in eleven blinks, some fifteen.

Given how quickly the heads were dropping—all twenty-eight men died within thirty-five minutes—the story is probably apocryphal. But as a myth it still holds a lot of power: it shows us a scientist dedicated to his craft even beyond his own death. It's as stirring as Socrates still doing philosophy while he sipped the hemlock. Indeed, had it happened, this would have been the greatest experiment of Lavoisier's life—entirely original, suitable for an audience, and involving nothing more than his prodigious brain and his beloved oxygen gas.

Interlude: Hotter Than the Dickens

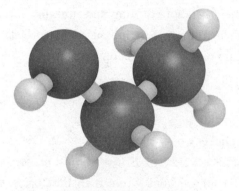

Ethanol (C_2H_5OH) — currently 0.00005 parts per million of air; you inhale 600 billion molecules every time you breathe

The first thing they noticed was the smell, like someone frying rancid lamb chops. The two men were sitting in their flat in central London and chatting uneasily about their midnight appointment with old, alcoholic Mr. Krook downstairs. But ominous signs kept interrupting them. Black soot swirled around the room like the devil's dandruff, staining one man's sleeves. When the other man leaned on the windowsill, his hands came away streaked with yellow grease. And that smell! Every minute they waited, the odor of rancid meat grew more and more overpowering.

Midnight finally struck, and they descended the stairs. Walking through Mr. Krook's shop—crammed with rags, bottles, bones, and other trash—was unpleasant even during the daytime. Tonight they sensed something positively evil. Outside Krook's bedroom near the back of the shop, a black cat leaped out and hissed. Grease stained the walls and ceiling inside the bedroom as if painted on. Krook's coat and cap lay on a chair; a bottle of gin sat on the table. But the only sign of life was the cat, still hissing. They swung their lantern around, searching for Krook.

They finally spotted a pile of ash on the floor. They stared stupidly for a moment—before turning and running. They burst onto the street and shouted for help, help! But it was too late. Old Krook was dead, a victim of spontaneous combustion.

———◄○►———

When Charles Dickens published this scene in December 1852—an excerpt from his novel *Bleak House*—most readers swallowed it completely. After all, Dickens wrote realistic stories, and he took great pains to depict scientific matters like smallpox infections and brain damage. So even though Krook was fictional, the public trusted that Dickens had portrayed spontaneous combustion accurately.

But a few members of the public couldn't read about Krook's death without themselves burning up, in rage. Scientists at the time were laboring to debunk old nonsense like clairvoyance, mesmerism, and the idea that people sometimes burst into flames for no reason. Within two weeks, skeptics began challenging Dickens in print, igniting one of the strangest controversies in literary history—a controversy about the role of oxygen in human metabolism.

Leading the charge against Dickens was George Lewes, a Victorian-era Richard Dawkins, always ready to attack superstition.

The death of Mr. Krook, from Charles Dickens's *Bleak House*.

Lewes had studied physiology as a young man, so he understood the body. He also had a foot in the literary world as a critic and playwright and as George Eliot's longtime lover. He counted Dickens as a friend.

Not that you'd know that from Lewes's response to *Bleak House*. He acknowledged that artists have a license to bend the truth sometimes, but he protested that novelists can't just ignore the laws of physics. "The[se] circumstances are beyond the limits of acceptable fiction," he wrote. He furthermore accused Dickens of cheap sensationalism and "of giving currency to a vulgar error."

Dickens swung back. He was publishing *Bleak House* in monthly installments, so he had time to slip a rebuttal into the January episode. The action picks up with the inquest into Krook's death, and Dickens mocks critics of spontaneous combustion as eggheads too blind to see plain evidence: "Some of these authorities (of course the wisest) hold...that the deceased had no business to die in the alleged manner," Dickens wrote. But common sense eventually triumphs, and the coroner in the story declares, "These are mysteries we can't account for!"

In private letters to Lewes, Dickens continued his campaign, mentioning several cases of spontaneous combustion throughout history. He leaned especially hard on the case of an Italian countess who combusted in 1731. She reportedly bathed in brandy to soften her skin, and the morning after one such bath her maid walked into her room to find the bed unslept on. As with Mr. Krook, soot hung suspended in the air, along with a yellow haze of oil. The maid found the countess's legs—just her legs—standing several feet from the bed. A pile of ashes sat between them, along with her charred skull. Nothing else seemed amiss, except for two melted candles nearby. A priest had recorded this tale, so Dickens considered it trustworthy.

Nor was he the only author to believe in spontaneous combustion. Mark Twain, Herman Melville, and Washington Irving all had characters erupt as well. As with the "nonfiction" accounts, most of these scenes involved old, sedentary alcoholics. Their torsos always burned to ash but their extremities often survived intact. Most eerie of all, beyond the occasional scorch mark on the floor, the flames never consumed anything but the victim's body.

Believe it or not, Dickens and these other authors had some science backing them up here. Dickens wrote *Bleak House* within a decade of the discovery of nitroglycerin—an explosive oil that can indeed detonate spontaneously. More significantly, spontaneous combustion seemed tied up with one of the most important discoveries in medical history, which linked the seemingly separate phenomena of burning, breathing, and the circulation of blood.

In 1628 William Harvey provided the first real evidence that blood flows around the body in a circuit, with the heart acting as a pump. (People before this assumed that the liver converted food into blood and that our organs "drank" blood the way plants do water.) At the same time, Harvey made some dubious guesses about the circulation of other fluids, like air. He knew that both blood and air passed through the lungs, but he insisted that the two fluids didn't mingle there. Instead, he argued that the lungs merely cooled the blood by churning it, in much the same way you'd stir soup to cool it. In other words, the lungs had a mechanical role but didn't alter blood chemically—only the heart could do that.

In the 1660s, Robert Hooke and Robert Lower—members of a new scientific boys' club called the Royal Society—finally disproved Harvey's theory about the lungs merely cooling the blood. They did so through a series of bloodcurdling experiments that involved vivisecting a dog. I'll spare you the worst details, but they snipped small holes in the dog's lungs to allow air to stream through, and slipped the nozzle of a bellows into its windpipe. Repeated pumping of the bellows kept the lungs inflated, like a windsock in a gale. As a result, the lungs remained stationary for several minutes at a time, neither expanding nor contracting.

The dog's heart and other organs worked just fine as long as air kept streaming through the lungs, despite their immobility. Contra Harvey, then, the mere *motion* of the lungs meant nothing. The duo also saw the dogs' blood change as it moved through their lungs, switching from a somber dark red to a bold Matisse red. All this lent strong support to their theory that the lungs did indeed trigger a chemical change in the blood, either infusing it with some substance or removing waste fumes.

Both, it turns out. As we saw earlier, chemists in the late 1700s determined that the lungs take in oxygen and expel carbon dioxide. (As for the color, when oxygen enters red blood cells it latches onto hemoglobin molecules there. Hemoglobin contains iron atoms, which bond readily to oxygen, and the addition of oxygen changes the shape of hemoglobin. This in turn changes its color from maroon to bright red.) In addition to connecting oxygen with breathing, these chemists had already linked oxygen with combustion, burning. So when they realized that blood delivers oxygen to our cells, they declared, QED, that breathing must involve a sort of slow combustion inside us—a constant burning, with our own bodies acting as the fuel.

And if slow fires burned inside us all the time, why couldn't they flare up on occasion, especially in alcoholics, whose very organs were dripping with gin or rum? To this way of thinking, spontaneous combustion didn't seem ridiculous at all. (Plus, not to put too fine a point on it, we all pass flammable gases several times each day.) As for what sets the fires off, perhaps fevers did, or raging hot tempers. In defending spontaneous combustion, Dickens was throwing fuel onto a smoldering scientific debate.

Lewes, however, was having none of this. He read over Dickens's historical accounts and dismissed them as "humorous but not convincing," noting that several were over a century old. It didn't help that Dickens enlisted the endorsement of a celebrity doctor who promoted phrenology as well. Lewes also pointed out that none of the "factual" accounts were written by eyewitnesses. The authors had always heard the story secondhand, from a cousin's friend or landlord's brother-in-law.

Most damning of all, Lewes had a better grasp on modern physiology. He pointed out recent work showing that the liver metabolizes alcohol, breaking it down for elimination, so despite what their breath might smell like, the organs of alcoholics aren't soaking in booze. Even if they were, the body is roughly three-quarters water, so it wouldn't catch fire anyway. And doctors knew by then that fevers don't burn nearly hot enough to ignite anything.

Not surprisingly, Dickens dug in. He'd always had an ambivalent relationship with science. He couldn't deny the marvels that science had wrought, but he was fundamentally a romantic and thought that science killed imagination. Artistically, too, he considered the scene with Krook so central to his novel (which involves a ruinous court case that "consumes" the lives and fortunes of everyone involved) that he couldn't stand it being picked apart. And the more defensive Dickens got, the more disgusted Lewes felt. They continued to bicker for ten months, then mutually dropped the matter when the final installment of *Bleak House* appeared in September 1853.

History of course has judged Lewes the winner here: outside the tabloids, no human being has ever spontaneously combusted. But back then the idea of spontaneous combustion wasn't as vulgar and ridiculous as Lewes claimed; one medical text was discussing cases as late as 1928. Plus, Dickens was undeniably right about one thing: that in human affairs, spontaneous combustion can happen. Dickens and Lewes eventually patched things up, but for those ten months in 1853 the fires burned awfully hot in London. They'd be the first to tell you that friendships and reputations can ignite instantly and consume themselves in smoke and ashes.

II. Harnessing Air

————◇————

THE HUMAN RELATIONSHIP WITH AIR

At this point, we've answered two major questions about air: where our atmosphere came from and what its major ingredients are, nitrogen and oxygen. Every time you inhale, though, you absorb a hundred other gases as well, gases that add overtones and finish to what we breathe, and our understanding of the atmosphere would remain superficial without their stories. These gases also present a new opportunity: rather than simply chronicle where each gas came from, we can now begin to examine the human *relationship* with air — how we've used these gases to improve our lives in hundreds of ways, beginning with medicine.

CHAPTER FOUR

The Wonder-Working Gas of Delight

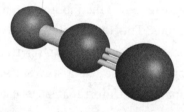

Nitrous oxide (N_2O)—currently 0.33 parts per million of air; you inhale four quadrillion molecules every time you breathe

One morning in 1791, an Oxford professor named Thomas Beddoes had the delicious pleasure of hearing two strangers gossiping about him. He was breakfasting at a stagecoach inn when he overheard his name, and upon sitting down he realized that neither of his tablemates recognized him. A mischievous sort, Beddoes egged them on, and smiled to hear the young man claim that Beddoes had just discovered three new volcanoes in England. The woman, although she granted that Beddoes was a scientific prodigy, scowled over his atheism and support of the French Revolution. "Excepting what he may know about fossils and such out-of-the-way things," she assured him, "he is perfectly stupid and incurably heterodox. Besides, he is so fat and short that he might do for a [freak] show." Beddoes slipped away with a chuckle.

The thing is, his companions didn't know the half of it. Over the years Beddoes would develop a reputation as the queerest man in English science. He exposed tuberculosis patients to cow farts to clear their lungs. He sucked on ingots of silver and lead to "taste" the nougat of electricity inside. And he preached these ideas to the biggest crowds that any professor at Oxford had enjoyed since the Middle Ages. Most notoriously, Beddoes promoted the use of mind-altering drugs like nitrous oxide (N_2O, laughing gas) to explore human consciousness.

Although most had been discovered just a few years earlier, several gases had already escaped the laboratory and become popular drugs by the late 1700s. This represented a new turn in pneumatic chemistry: rather than just measure the properties of different airs, as earlier generations had, these scientists were working with an eye toward helping humankind. Unfortunately, as usually happens with medicine, the field attracted charlatans as well. Depending on which quack you credited, different gases could cure typhus, ulcers, and diabetes; croup, catarrh, and pleurisy; diarrhea, scurvy, and sore throats; even blindness and deafness. But no gas stirred up as much excitement, and dread, as nitrous oxide. Beddoes's work with this gas would straddle an uncomfortable line between science and crankdom. And although his experiments with laughing gas would lead, in time, to one of the great breakthroughs in medical history, he died thinking himself a failure.

Beddoes trained as a doctor, and he practiced medicine the way Socrates practiced philosophy, criticizing everyone, without regard to rank or station. He excoriated gentleman doctors for ignoring the suffering masses, but he denounced with equal fervor those who

Thomas Beddoes, medical gadfly.

preyed upon the masses by selling useless balms and tinctures. Beddoes pushed for a middle path, and he suddenly realized how he could achieve this while tramping through a muddy pasture one day in 1791.

Most doctors in that era traced the origins of every disease to poisoned pockets of air. (Malaria literally means "bad air.") This explains why sickly men and women in old novels were always flocking to seaside resorts and mountain sanatoriums, places where they could breathe free and easy. Beddoes believed in good and evil airs as much as anyone, but he'd also studied chemistry under Joseph Black, discoverer of carbon dioxide. While taking a mud-spattered constitutional one day, an idea struck him: Why not manufacture his own airs and force patients to breathe them?

Inspired, Beddoes began collecting case reports of people who'd exposed themselves to different gases; he also began conducting experiments on himself. He found that huffing oxygen, for instance, rendered him immune to cold and melted fifteen pounds off his frame in just a few weeks. (Unfortunately it also dried out his skin and caused alarmingly bright-red nosebleeds.) He eventually collected these experiments and reports into a wide-ranging book that included several dubious claims—that gases could shrink tumors, for instance, and abolish the need for sleep.

Meanwhile Beddoes sketched out plans to open a research center, the Pneumatic Institution, where he could test gases systematically. (He was about to be fired from Oxford anyway for writing handbills supporting the French Revolution.) He envisioned the Pneumatic Institution as both a hospital for treating patients and a laboratory for testing new cures—perhaps the world's first true medical research center. The venture wouldn't be cheap, so Beddoes sought patrons among Joseph Priestley's old social club, the Lunar Society of Birmingham. Several eminent members ponied up, including the Wedgwoods of pottery fame; James Watt of steam engine renown agreed to build equipment for him at cost.

Generosity aside, Watt had a personal stake in Beddoes's work. Among other diseases, Beddoes planned to target consumption (tuberculosis of the lungs). Like gas warfare victims, people suffering from consumption slowly drowned as their chests filled with fluid. They also suffered from chills and sweats, and hacked up blood while they waited to die. Watt knew these symptoms all too well because his daughter Jessie had consumption. She'd already tried an endless number of remedies—foxglove, laudanum, bark tea, blistering, bleeding, even "being swung about on a rope to make her sick." As a last-ditch effort, Watt let Beddoes treat her with a gas, carbon dioxide. She died within a week.

Beddoes cringed, fearing Watt's wrath. Famous as he was, Watt could have destroyed the Pneumatic Institution before it even opened its doors. But Watt was a kind and reflective man, and rather than blame Beddoes he redoubled his commitment. To that end Watt invented a portable furnace with various distillation tubes and reaction chambers to create new gases on demand. He also developed clever ways to collect gases, either in bellows (for pumping into the lungs) or green silk bags with mouthpieces (for huffing at leisure). The whole arrangement, which cost just £14, delighted Beddoes, who declared that preparing doses of gases would soon be "as easy as...dress[ing] a joint of meat."

Funds and equipment in hand, Beddoes next focused on securing land and a capable assistant. For the grounds he chose Bristol, an inexpensive city with natural springs that attracted flocks of consumptive patients. Indeed, one historian noted that Bristol "had become a 'last-chance saloon,' the grim terminus for those for whom all other treatments had failed....The proprietors of guesthouses and hotels often doubled as funeral directors." Beddoes calculated, correctly, that many patients would be desperate enough to try his gas cure.

For an assistant, Beddoes once again turned to Watt. In the late 1790s, Watt's son Gregory—also consumptive—took a rest leave in southwest England and lodged with a widow named Grace Davy. The widow's teenage son Humphry had been earning quite a reputation locally as both a brilliant chemist and a Beddoes-level eccentric. He once built an air pump—a sophisticated piece of equipment then—from an enema syringe that washed ashore after a shipwreck. He painted goblins on the walls of his sisters' room with glow-in-the-dark phosphorus. He wrote intense, visionary poems and took long, lonely walks along the cliffs of Cornwall, often returning home bruised and bleeding.

Although repelled at first, Gregory Watt warmed to Davy as the months wore on. They became drinking buddies (mostly brandy), and he encouraged Davy to write to Beddoes. Excited to have contact with a real-life scientist, Davy sent Beddoes two hundred pages of rambling essays about heat, light, electricity, and gases. Most had nothing to do with medicine, but they revealed a sharp scientific mind, and a few months later Beddoes hired Davy—whom he'd never met and who was just nineteen years old—to run experiments

Humphry Davy, chemist, romantic poet, and element-hunter.

at the Pneumatic Institution. Davy had never strayed more than a day's travel from home before, but in October 1798 he made the two-hundred-mile trip to Bristol; to save money, he bought the cheapest ticket available, and had to ride on top of his coach. His first impression of Beddoes was that he was short and fat.

The clinic opened in March 1799, and by early April, Davy had almost killed himself. Most of his day-to-day work involved preparing gases and measuring their chemical properties. He also exposed dogs, cats, rabbits, and butterflies to gases and monitored how their breathing and heart rates changed. But above all Davy wanted to inhale gases himself, and for his first full-blown experiment he prepared several quarts of carbon monoxide. After the third hit, his pulse began racing and his chest seized up. He barely managed to stumble into the garden, where a terrified assistant revived him with oxygen. Davy spent the rest of the day in bed vomiting and suffering an excruciating headache.

No matter. After a week off Davy jumped right back into his studies, this time focusing on another supposedly poisonous gas, nitrous oxide. He prepared nitrous by heating crystals of ammonium nitrate in a sealed vessel—slowly, to prevent an explosion. He then collected the fumes in a bellows. His first impression was that it tasted sweet. A few more puffs left him dizzy, but his hearing grew acute. Then he noticed an odd tactile sensation, "a gentle pressure on all muscles." The episode climaxed with him leaping up and marching about the room, shouting for joy. Beddoes, observing all this, said that Davy appeared to be having a "higher orgasm," and Davy could hardly sleep that night for all his racing thoughts.

After a few more weeks of testing, Davy and Beddoes had proved that nitrous wasn't poisonous, and they felt confident enough to try it on two patients. One had a half-paralyzed arm from an epic drinking binge a few years before. A few hits of gas put him right, unfreezing his hand and allowing him to grasp objects. The second patient was in worse shape—"as shattered a human creature as can easily be met with," remembered Beddoes. But he too responded to nitrous, rising like Lazarus and tossing aside his crutches.

Word about this remarkable gas started to spread through town, and it wasn't long before Bristol's bohemian set asked Beddoes and Davy if they could try it, too. Beddoes readily agreed—the psychonautic properties of the new gas fascinated him—and he encouraged several poets that Davy had befriended to come by for a nip of nitrous. The poets had a grand time that evening and returned for more another night. Then another. Pretty soon the Pneumatic Institution was leading a double life. By day it was a respectable clinic, with Beddoes treating patients and Davy running experiments. By night it resembled an opium den, with writers and their groupies lounging around and huffing nitrous gas from green silk bags.

Davy couldn't resist slipping a little science into these sessions. He tested people's sensory responses by having them follow candle flames or listen to dinging bells. He tested the power of suggestion by preparing placebo bags of plain air to see whether people still pretended to feel high. (Not really.) But above all, he recorded people's personal, individual responses to laughing gas. Some got rowdy or declaimed nonsense. One woman raced outside and, to her later mortification, hurdled a large dog. More often people slumped onto the floor and disintegrated into giggles. Later, Davy encouraged them to describe their sensations in words. Local writer Samuel Taylor Coleridge compared a nitrous high to entering a fire-warmed room after a snowstorm. Fellow poet Robert Southey gushed in a letter to a friend, "Such a gas Davy has discovered! . . . It makes one strong. And so happy! So gloriously happy! O, excellent gas bag! I am sure the air in heaven must be [made of] this wonder-working gas of delight!" (Clearly, one symptom of a nitrous overdose is abusing exclamation points.) Most eloquent of all, one man emerged from his high and said simply, "I felt like the sound of a harp." Rather than dismiss these sentiments as unscientific, Davy analyzed them for clues about the human psyche. Laughing gas seemed to open up new vistas within people's minds, and he needed poets and their linguistic gifts to capture all the subtleties.

Excited by this work, Davy began pulling fourteen-hour shifts. He rarely bothered eating square meals during them, and when his shirt became too filthy for decent company, he'd pull a clean garment over the top and keep going. (He did the same with socks.) Then, to unwind after a long day, he'd whip up a half-dozen bags of nitrous and get high. In fact he became something of a nitrous addict, huffing the gas daily for several months. Some nights he wandered the countryside and passed out under the full moon. Other nights he stayed in and mixed drugs. He once tried chugging a bottle of wine as fast as possible and then taking hits of gas. He ended up puking.

Another night he tested a new gas-immersion unit from James Watt. It consisted of a large box with a sedan chair inside, and Davy entered it half naked, a thermometer tucked in his armpit and a feather fan in his hand to stir the air inside. Over the course of seventy-five minutes an assistant released three hundred quarts of nitrous into the chamber. Davy emerged wobbly—flushed red and running a temperature of 106°F—but he insisted on taking one more hit from a silk bag. This rocketed him to the highest high of his life, a new dimension of intoxication. At one point he declared himself "a sublime being, newly created and superior to other mortals." Moments later, he babbled, "Nothing exists but thoughts! The world is composed of impressions, ideas, pleasures, and pains!" He sounded like a deranged Bishop Berkeley. But Davy considered such testimony just as valuable as his observations on heart rate or pupil dilation.

Beddoes mostly left Davy alone to conduct these experiments. That's partly because Beddoes was less interested in getting high, and partly because he had his own screwball project to pursue. Several years before, Beddoes had noticed that butchers never seemed to get consumption. In fact they seemed like the opposite of wan and weak consumptives—hearty and hale blokes with healthy figures. Intrigued, he asked around and learned that most butchers credited their health to the fumes they inhaled while carving up cows and sheep.

However ridiculous this sounds today, the idea struck Beddoes as plausible. A few years earlier his countryman Edward Jenner had noticed that milkmaids infected with pus from a bovine disease called cowpox never came down with the far deadlier smallpox. That insight led Jenner to develop a smallpox vaccine. So why couldn't cow vapors have medicinal effects as well?

To test the idea Beddoes outfitted a cattle shed with beds and moved a few consumptives in. At first he let the beasts wander where they would, farting and belching at will. "Living with cows is the most delicious thing imaginable," he assured his patients. His patients disagreed. Revolted at the sty they were living in, they insisted that Beddoes clean up the dung and install curtains. They nevertheless agreed to spend several months in the shed breathing cow exhaust.

Cartoon mocking Beddoes and Davy's gas experiments. (Photo courtesy Wellcome Trust)

However ineffective, this "cow-house therapy" was probably harmless from a medical standpoint, but it did end up damaging Beddoes's reputation—or rather, damaging it further. Beddoes's atheism and radical politics had already aroused suspicion around Bristol. (At one point Davy had ordered frogs to experiment on, and when their crate cracked open and they escaped into town, rumors began to swirl that Beddoes had bought them to feed French spies hiding in his basement.) Cow-house therapy gave his enemies yet another switch to thrash him with, and like Joseph Priestley before him, the short and fat Beddoes became a popular subject for caricature. The nightly nitrous binges provided even more fodder for satire, and Beddoes's whole research program soon became an object of mockery across Europe.

What ultimately doomed the Pneumatic Institution, however, wasn't politics or satire but science. Because while patients sure enjoyed the nitrous highs, very few of them actually got better. Again, a comparison to Jenner here is instructive. On the face of it, Jenner's idea of infecting people with pus from cowpox sores seems dubious, even dangerous, and tabloids attacked Jenner even more fiercely than they did Beddoes. (One story claimed that a local boy metamorphosed into a cow—grew horns and everything—after being vaccinated. However ignorant, today's anti-vaxxers have nothing on the know-nothings of yesteryear.) Still, no amount of ridicule from the smart-aleck set could change the fact that vaccines worked, and patients lined up by the thousands. Nitrous, meanwhile, cured no one. Even those who felt temporary relief from their symptoms soon built up a tolerance and sunk back into illness. Beddoes and Davy also saw an increasing number of bad reactions—patients with headaches, lethargy, malaise. One woman had "a succession of hysterical fits" lasting for weeks.

Despite these setbacks, Beddoes continued to promote his cures, never wavering in his belief that gases would somehow transform medicine. Little did he know, his sidekick Davy would soon provide his enemies with all the ammunition they'd need—and worse, would soon switch sides himself.

In his studies Davy could see that gases caused physiological changes, certainly. But the dearth of actual cures frustrated him, and he lacked Beddoes's unflagging optimism. He finally made his suspicions public in a six-hundred-page book he published in 1800, *Researches, Chemical and Philosophical.* Per the title, it's an erratic tome, wandering from straightforward chemistry experiments to Samuel Taylor Coleridge's thoughts about getting stoned. Its conclusion, though, was stark and concise: that gases were worthless as medicine.

Depending on your perspective, the book was either a betrayal of his mentor or a hardheaded but necessary corrective. Regardless, it pained Beddoes, and he and Davy soon split. On the strength of *Researches,* Davy landed a plum scientific job in London, where he expanded his research into new topics. For instance, he harnessed the power of electricity to liberate a half-dozen new chemical elements (sodium,* potassium, barium, magnesium, calcium, and strontium), a world record that would stand for 150 years. By the 1820s Davy had become the most famous scientist in England; he never could have dined incognito in a coaching inn while strangers gossiped about him.

Meanwhile, Beddoes's reputation tanked. Critics never left off mocking him, and when donations dried up in 1802, the Pneumatic Institution folded. It had lasted just three years, and Beddoes himself—once so lighthearted—died on Christmas Eve six years after that, still bitter over his failure. Indeed, he and his institute probably would have been lost to history if not for one thing.

As a teenager Davy had spent many hours rambling along the cliffs of Cornwall and had often returned home with scrapes and gaping wounds. He proved no less reckless in the Bristol lab—more than once he sliced his finger to the bone. He noticed, however, that the wounds stopped smarting as soon as he took a few hits of laughing gas. Toothaches and headaches also evaporated. Busy with other work, Davy didn't pursue this line of research, but he did include a passage about it on page 556 of *Researches*—a throw-away line that would eventually become the most famous thing he ever wrote. "As nitrous oxide...appears capable of destroying physical pain," he noted, "it may probably be used with great advantage during surgical operations." It took another half century, but this idea—anesthesia—would eventually resurrect the reputation of the Pneumatic Institution.

———◁○▷———

The story of anesthesia is the story of a swindler and his mark. No confidence man in history did more for humanity than William Morton, but he never could have revolutionized medicine without the hapless Horace Wells.

Before 1850 people routinely committed suicide rather than face surgery, and you can't really blame them. Every detail of the operating theater promised pain, from the floor—strewn with sand to absorb blood and vomit—to the ceiling—skylighted to let screams escape. At eye level patients were confronted with trays of picks and saws, not to mention rows of surgical gowns scabbed with blood. And once the butchery started, the main concern wasn't kindness or delicacy but speed. Operations more complicated than carving out a bladder stone or hacking off a leg simply weren't possible.

Doctors didn't enjoy surgery, either. A young Charles Darwin quit medicine forever after witnessing an operation on a howling boy, and even experienced surgeons might admit relief when patients flew the coop. With all the bucking and kicking and sharp blades involved, surgeons occasionally got killed, too. One observer marveled, "I don't wonder that the patient sometimes dies, but that the surgeon ever lives."

Today it seems obvious that laughing gas could eliminate many of these problems, but it didn't gain traction as an anesthetic for several reasons. Beddoes had made so many wild promises about gases that any suggestions about killing pain seemed like more hyperbole. And knocking people out with nitrous actually takes some skill. Low doses can in fact rev people up—the last thing surgeons wanted.

That doesn't mean that people ignored nitrous—to the contrary. After hearing Coleridge and other poets wax mystical about it, the public clamored for it, and laughing gas became a fad drug in Europe and North America. Traveling salesmen sold snorts for a quarter on street corners; rich folks served it in lieu of wine at dinner parties. (Chemists even discovered nitrous in the atmosphere— a nanobuzz in every breath!) Most commonly people encountered the gas in traveling stage shows, where volunteers would take several hits and then begin singing, dancing, or doing gymnastics for the audience's entertainment.

One evening in December 1844, a boyish-looking dentist named Horace Wells attended a nitrous frolic in Hartford, Connecticut, with a pal. After a few hits they began careening around onstage, and their minds went blank. Wells came back to his senses several minutes later—and was startled to see his friend's leg soaked in blood. The friend was equally startled; he had no idea what had happened. (Witnesses later said he plowed into a couch.) Even more surprising, the friend realized that he was in excruciating pain, but until that moment he'd noticed nothing—not until his high faded.

That night, Wells chewed and chewed on what his friend had said—*I felt no pain until it was over.* The very next morning he hunted down the show's emcee and dragged him to his office, along

Horace Wells

From the Engraving by H. B. Hall

Horace Wells, anesthesia pioneer and hapless businessman.

with another dentist. The emcee prepared a big bag of nitrous, and Wells huffed it and passed out in his dental chair, head lolling back. Working quickly, the dentist friend grabbed some pliers and yanked out a wisdom tooth that had been bothering Wells. Wells came around several minutes later and felt the empty socket with his tongue. "I didn't feel so much as the prick of a pin," he marveled.

Over the next few weeks, Wells tested laughing gas here and there in Hartford, and it seemed promising. But he knew that the real challenge would be winning over Boston, the country's leading medical center. So he got in touch with an old business partner there, William Morton, a through-and-through scoundrel.

After quitting school as a teenager, Morton had worked at a tavern in Worcester, Massachusetts, where he got caught stealing from the till. Over the years he progressed to passing bad checks, embezzlement, and mail fraud. He also jilted several fiancées and got excommunicated from his church. Rochester, Cincinnati, St. Louis, Baltimore—he was run out of nearly every major city in America. Still, his good looks, charm, and natty suits ensured him a warm welcome wherever he drifted next.

**William Morton, anesthesia pioneer
and successful confidence man.**

Morton finally decided to make an honest living and apprenticed himself to Wells. He found he wasn't bad at dentistry: dentists needed little medical training then, and confidence and presentation—

Morton's forte—went a long way. He soon opened his own practice and married a nice girl. But when Wells invented a new type of gold platework and proposed that he and Morton go into business, the old itch for a quick buck overwhelmed him. Morton swiped the capital that Wells had raised and spent it on himself.

Wells must have been pretty desperate for contacts in Boston to get in touch with Morton after that. Morton, meanwhile, was thrilled to be let in on this killer idea. Thousands of patients in the United States and Europe had teeth yanked out every day, and Morton had visions of peddling nitrous to every last clinic. So in January 1845—before Wells felt ready—Morton arranged for a public demonstration at Massachusetts General Hospital.

The operating theater at Mass General was cozy, a small amphitheater with rows of wooden benches for spectators. A mummy stood in one corner, and the wall behind the operating floor was studded with hooks, rings, and pulleys—for holding patients down with ropes.

Wells could have made those hooks and pulleys instantly obsolete, but fate had other plans. The original patient had hobbled off in fright before Wells arrived, so a medical student in the crowd volunteered to have a nagging tooth extracted. (Seemingly everyone had a bad molar or two back then.) With Morton watching from the audience, Wells knocked the student out and began to wriggle the tooth. As for what happened next, theories differ. Perhaps Wells hadn't prepared enough gas. (Young people generally need more anesthesia because their livers metabolize it quicker.) Or perhaps Wells drew the bag away from the student's lips too soon. Or perhaps he did nothing wrong—some people just don't go under very deeply. Regardless, when Wells tugged the tooth, the student groaned. Later, when the student came around, he insisted he hadn't felt anything; the gas had worked. But it was too late. As soon as the audience heard that groan they began shouting "Humbug! Humbug!" Even William Morton had never been run out of town so fast.

Morton only shrugged at the setback. What were some loud-mouthed medicos compared to the Baltimore police or a jilted debutante's father? If nitrous didn't work as an anesthetic, well, he'd find something better. And the darn thing was, he did. Despite a complete lack of chemical or medical training, within a year he'd hit upon ether.

At a glance, ether doesn't seem promising as an anesthetic: you generally inhale anesthetics, and ether is a liquid at room temperature (it boils at 94°F). What saves ether is its tendency to evaporate, its volatility. Volatility in a liquid depends on two factors, weight and polarity. Weight is easy to understand. Evaporation occurs when the molecules on the surface of a liquid vault upward and become gaseous, and it's easier for a molecule to become airborne when it weighs less. (There's a reason gymnasts are tiny.) Polarity means electric charge. Atoms are generally electrically neutral, but when they start swapping electrons and arranging themselves into molecules, they can become charged. Oxygen, for example, usually steals electrons from its neighbors and acquires a negative charge. Hydrogen, meanwhile, often gives up its electron and drifts positive. Polar molecules, like H_2O, have both negative and positive regions. As a result they're less volatile and they stay liquid longer than nonpolar molecules, because the positive and negative ends attract each other like magnets. This attraction makes it harder for polar molecules to leap up and become gaseous.

Every substance has a different volatility based on its weight and polarity. Nonpolar gases like nitrogen and oxygen are maddeningly volatile; each boils at around −300°F. In contrast, the highly polar water molecule—despite weighing less than nitrogen or oxygen—remains liquid at temperatures five hundred degrees higher. Ether $(C_2H_5-O-C_2H_5)$ falls in between. It's heavy, four times the weight of water. But it's only barely polar: those ten hydrogens are spread fairly evenly around the molecule, making its surface mostly positive. As a result, ether molecules show little affinity for one another: if you put a glass of water and a glass of ether on your countertop, the ether would evaporate dozens of times quicker. That's useful for an anesthetic.

Morton was acquainted with exactly none of this science. But he was acquainted with ether, a cheap high* that dulled sensation, the marijuana of its day. So he began testing ether on the animals on his father's farm: cows, horses, worms, the family spaniel, even a few fish.* These trials went well, so he purchased another batch of ether (from a different druggist, to mask his plans) and put a friend with an impacted wisdom tooth under. Another success: the friend woke up sans tooth, wondering when the procedure would start. Dollar signs dancing before his eyes, Morton once again raced off to Mass General to arrange a demonstration in October 1846.

He made those arrangements with Dr. John Warren, in some ways the hero of this story. Although eccentric—he spent much of his free time reconstructing a mastodon skeleton—Warren was the most eminent surgeon in America, and at age sixty-eight he could have coasted into retirement. But the pain that he inflicted on his patients had always haunted Warren. True, a few painkillers did exist then. Patients could drink themselves into stupors or smoke opium. They could numb a limb with ice or let doctors bleed them until they passed out. Some doctors also practiced "concussion anesthesia": they'd swaddle the patient's head in a leather helmet, then smack his skull with a mallet. (If that failed, there was always a haymaker to the jaw.) But each approach had downsides, even beyond the obvious. Alcohol thinned the blood, for instance, making hemorrhages more likely. And no approach could reliably put patients under or obliterate the memory of surgery.

Most surgeons simply sighed and accepted pain as inevitable, one of the evils of life. Warren, though, fended off such cynicism. He was past the age at which most men make great discoveries, but he held out hope that he could somehow contribute to making anesthesia a reality. So when Morton approached him about ether, Warren swallowed his misgivings. A local house painter needed a tumor beneath his left jawbone removed, and Warren told Morton to report at 10 a.m. sharp on Friday, October 16.

Morton spent that morning in an uncharacteristic panic. While certainly not opposed to ending suffering worldwide, he was pursuing anesthesia mostly to cash in. Problem was, ether was a common chemical, which meant he couldn't patent it. Morton therefore planned to keep its identity secret. But the same property that made ether a viable anesthetic, its volatility, threatened to screw him here. Ether had a sickly sweet aroma, and the fact that it evaporates so readily ensured that this wasn't a subtle aroma. Ether reeked. Morton tried disguising its smell with orange zest, but the ether always broke through. So Morton—who until then had just dumped ether on a cloth for patients to inhale—constructed a special breathing apparatus to bolster his chances of securing a patent.

Unsatisfied with the device, Morton spent the day and night before the demonstration designing a new one—or, rather, conning some mechanically adept friends into designing one. He then rushed off at dawn to a machinist to throw something together. The result consisted of a glass bulb, a sponge inside the bulb to hold ether, and various valves and pipes to allow air to flow through. All in all it looked like a hash pipe. Because it took so long to finish— ten o'clock had already struck by the time Morton raced out the door—he arrived at Mass General without having tested the thing.

Think about the chutzpah here. A man with no medical training, who'd been awake much of the night, was going to administer a largely untested drug from a device he'd never used before, while the most prominent surgeon in the country watched. Even Morton's ever-loving wife doubted him, and she spent all day pacing back and forth in their kitchen, convinced that Morton would kill the patient and land in jail. But as soon as Morton had the inhaler in hand, he felt that old con man's swell of confidence, and a smile returned to his lips.

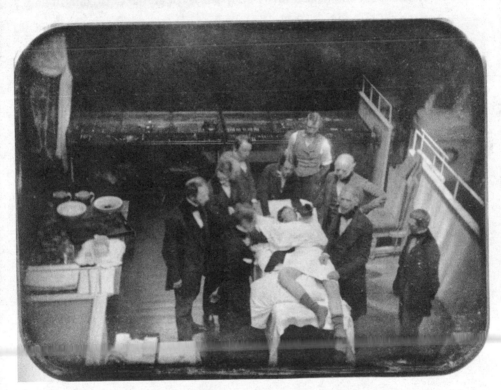

Reenactment of the first surgery using ether anesthesia.

Meanwhile, Dr. Warren was seething. The audience had already gathered in the operating theater; the house painter lay ready in a smock and socks. But 10:10 came and went with no Morton. Then 10:20. The mummy in the corner stared. Finally Warren picked up his scalpel, sighed, and turned to the patient. Another hope dashed. The tumor below the man's jaw was basically a swollen vein, red and blotchy, and it was so large that it bulged upward into his mouth—the patient could feel it with his tongue. The removal would not be quick; it would be painful. The house painter leaned back and gripped the bed, ready to buffet the ceiling with screams.

At that very moment Morton bounded through the door, charm wafting off him like ether itself. His entire life as a confidence man had prepared him for this moment. Warren turned and said, bitingly, "Your patient is ready." Morton smiled and leaned in with the ether bong, working the valves like a pro. A few minutes later the house painter drifted off, submerged in oblivion. Morton wheeled on Warren. "*Your* patient is ready," he said.

The surgery took longer than expected, but otherwise went perfectly; the house painter never stirred. And the audience that day— mostly med students, who'd arrived having no idea what was in store—later remembered one thing above all. The silence. No screaming, no thrashing, just the silken sound of a scalpel cutting flesh. Warren and Morton had traded barbs that morning, but after the surgery wrapped up, it was Warren's final words that would go down in history. "Gentlemen," he announced, "this is no humbug." The audience cheered.

———◁◦▷———

Surgeons soon learned that ether had several advantages over nitrous, in that it produced a deeper unconsciousness and required less skill to get the dosage right. This in turn allowed them to develop longer, more complicated procedures that reached deeper inside the body. In a wider sense, anesthesia also helped salvage the very reputation of surgery. For centuries other doctors had scorned surgeons as butchers—an unfair but understandable judgment. Anesthesia reversed that verdict, made surgery seem heroic.

Even a rogue like Morton was awed by the humanitarian potential of ether. On the day he proved ether's worth, he finally returned home at four o'clock and found his wife still pacing, still convinced that Morton would end up in jail. The preoccupied look on his face seemed to confirm her worst fears, but deep down, Morton was reflecting on what had just happened — on the new era in medicine that had just opened. His usual glibness absent, he embraced his wife and said simply, "Dear, I succeeded."

What he didn't succeed at was making any money off his discovery. Over the next year he managed to secure a patent on the general idea of chemical anesthesia, but even a D+ chemist could recognize the odor of ether, and word soon got out to doctors and dentists. After that, probably no patent in history was violated so widely and with so little consequence. Morton might as well have tried to patent water. A low point came when American army surgeons began using ether during the Mexican-American War without compensating Morton — meaning the U.S. government was violating its own patent.

Facing ruin, Morton began clamoring for a $100,000 bounty from Congress, a lump payment in lieu of lost revenue. Despite several champions in the House and Senate, these efforts came to naught. For one thing, Morton's grubbiness and dishonesty repelled even most politicians. (As the years passed, Morton would claim to have personally administered ether to 200,000 Civil War soldiers, a ridiculous claim.) Congress also declined to pay him because rival claims popped up* — people who swore they'd come up with the idea of anesthesia years before Morton had. A doctor in Georgia, Crawford Long, had used ether in surgery in 1842, knocking a man out with an ether-soaked towel and removing two tumors from his back; he also amputated a slave boy's toe. Given these cases, Congress decided that Morton had done nothing special. He kept fighting for his $100,000, but he finally suffered a stroke and lost his mind in

July 1868 while riding in a New York City carriage. (He demanded that the driver stop in Central Park and then dove into a lake there to "cool off.") He died in an insane asylum days later, the debate about his legacy still raging.

With several other claims out there—many of which hold up historically—why does Morton get credit for discovering anesthesia today? Was this simply his last, greatest con—hoodwinking history? Not quite. A dozen people before Morton had *proposed* using anesthesia, sure. So what? Morton actually tried it on patients, and it would have been his ass in a sling if he'd failed. Morton deserves more credit than Wells because, luck of the devil, his compound worked and Wells's didn't. If you want to credit Crawford Long over Morton, that's defensible, but science isn't a private enterprise. Science is public, and in some sense scientific discoveries don't count until they're public. For various reasons—timidity, lack of confidence, accusations of sorcery from his neighbors in Georgia (really)—Long hoarded his ideas, and thousands of people suffered meanwhile. In the end Morton does come off as distasteful; almost everyone who got to know him regretted it. But he had the guts and/or irrational confidence to pursue anesthesia. And although he himself died unhappy, if you're into utilitarianism— the greatest happiness for the greatest number—then William Morton, in making anesthesia a medical reality, did more to benefit humankind than almost anyone who ever lived.

———◦———

Morton's sometime partner Horace Wells died in circumstances even more pitiable than Morton did—destroyed, ironically enough, by the century's other great anesthetic.

Chloroform found its way into medicine because of Dr. James

Simpson, a trollish-looking obstetrician in Scotland who wanted to ease the pain of childbirth but found ether unsuitable: It worked too slowly, the smell made pregnant women vomit, and it required heavy doses before it took effect. So on his evenings off Simpson began hunting around for a substitute. His experiments didn't exactly adhere to the rigors of the scientific method: he'd dribble random chemicals into a tumbler of hot water, then sniff the fumes until he felt woozy. One historian described his approach as more like "the exploits of a teenage glue sniffer" than a proper scientist; a friend habitually stopped by at breakfast time to see if Simpson had killed himself during the night. But in 1847 Simpson hit upon chloroform, $CHCl_3$. Like ether, chloroform is only somewhat polar—as a molecule, most of its external facets have a negative charge. But because it weighs roughly double what ether does, chloroform is more sluggish and evaporates less quickly, allowing doctors to better control how much patients inhaled.

However useful for surgery, chloroform proved controversial within Simpson's field. Some obstetricians argued that mothers wouldn't bond with their children unless they'd experienced enough pain first. Others worried—on what evidence, it isn't clear—that anesthesia would somehow convert pain into pleasure, as if labor would suddenly become a giant orgasm. Many also feared contravening the will of God, who had cursed Eve with the pain of childbirth for bringing sin into the world. A few physicians even suggested withholding baptism from children born under anesthesia. Simpson countered such objections by quoting Genesis right back: when creating Eve, "the Lord God caused a deep sleep to fall upon Adam" before He removed Adam's rib—a clear reference to anesthesia. The argument was finally settled when John Snow (the cholera doctor of "ghost map" fame) administered chloroform to Queen Victoria during the birth of her seventh child,

Prince Leopold, in 1853. After that, doctors couldn't praise chloroform fast enough.

Unfortunately, like ether and nitrous before it, chloroform proved addictive, and probably the best-known addict was Horace Wells. After his disgrace at Mass General, Wells had quit dentistry to pursue other lines of work—peddling showerheads, smuggling exotic birds, selling knock-off paintings to rich rubes in the United States. The latter scheme brought him to Paris where, to his shock, medical men revered him as a genius for his early work in anesthesia. King Louis-Philippe invited Wells to serve as the royal dentist, an honor he declined.

After returning to the United States in 1847, Wells relocated, sans wife and son, to Manhattan. His trip to France had rekindled his interest in anesthesia, and he acquired some chloroform and began testing it on himself. Alas, his time in France had not lessened the pain of his humiliation, and he began dissolving himself daily in a chloroform haze.

Addiction, and the need to feed it, soon brought Wells into contact with hooligans, which proved his undoing. One thug burst into Wells's apartment in January 1848 complaining that a prostitute had just thrown vitriol (sulfuric acid) on him, ruining his cloak. He asked Wells to mix up a vial of vitriol so he could retaliate. To the strung-out Wells this sounded only fair; he decided to accompany his friend. The avengers succeeded—they ruined the woman's dress—but when the friend suggested extending the escapade and targeting other women, Wells demurred.

The idea stuck in his mind, however, and during a chloroform hallucination a few days later, he snatched the vial off the mantel and rushed down to the street. There, he doused two prostitutes in vitriol, burning one's dress and leaving the other's neck sizzling. Wells ran on, deranged, and lost all memory of what happened next.

Two other ladies of the night eventually disarmed and detained him, and they told the cops that they'd seen Wells splash acid into a young woman's face, sending her to the hospital and scarring her for life.

A judge set an enormous bail, but did allow Wells to return home under police escort to collect some toiletries. While the attending officer was distracted, Wells ducked into his bathroom and secured a razor blade and one last vial of chloroform. The next night, after attending Sunday services in prison, Wells wrote a letter to his wife in Connecticut, who had no idea he'd been arrested. His affairs in order, he doused a handkerchief in chloroform, stuffed it into his mouth, and secured it with another handkerchief. As the anesthesia took hold, he hacked into his thigh with the razor and severed his femoral artery. Guards found him dead the next morning in a sticky red pool.

Later that day, a policeman decided to seek out the poor woman whose face had been splashed with acid and tell her of Wells's death, to give her some closure. The nearest hospital had no one who fit her description, so he tried another. Then another, and another. No one had any idea what he was talking about. It later came out that the prostitutes had made up the story. It was one last squalid detail in a story overburdened with them. Wells had figured out how to abolish pain for the entire world, but he never could conquer his own anguish.

———<o>———

Nitrous oxide, ether, and chloroform were all single drugs, single compounds. Nowadays anesthesia usually consists of a cocktail of several drugs, each of which targets a different physiological function. Some of them slow breathing down, some paralyze muscles;

others relieve anxiety or interfere with memory formation. So in one sense we know a lot about how these drugs work, since we can measure exactly how much they affect blood pressure and body temperature and a dozen other signs.

In a larger sense, though, we know zilch about how these drugs work, since we don't know how they affect the brain. That's a little scary. We do know that anesthetic compounds dissolve preferentially into fatty brain tissue, and they obviously interfere with neuron function somehow. Beyond that, umm...The problem is that anesthesia disrupts consciousness—hits pause on it, essentially—and we have only a vague idea of how consciousness works in the first place.

But several recent studies on anesthesia have peeled back some of the mystery. One surprise is that the brain doesn't just shut off under the influence of anesthesia. Consider a person lying on an operating table under heavy sedation. If a surgeon cuts something and says "Whoops!," her eardrums still capture the noise and the auditory parts of her brain still crackle with activity. Same with smells: if the surgeon forgot to wear deodorant, the olfactory centers in her brain still register this. Even when sedated, we aren't oblivious to the world around us.

That said, anesthesia does interfere with the next steps in cognition. In a wide-awake person, the sounds and smells would now zip off to other parts of the brain and stimulate a response—*uh-oh,* or *ewww.* Under sedation, these signals never get the chance; they flat-line, and the rest of the brain never hears about them. (As a neuroscientist might say, her brain has received but not *perceived* these signals.) In other words, while anesthesia doesn't shut down the brain entirely, it does hush the chatter between different parts of it.

These studies also shed light on how people emerge from anesthesia. Intuitively, you might think that anesthesia simply "wears off" and that you emerge from the deep at a steady rate, but no. Rather, the brain seems to quantum-leap from one slightly-less-with-it stage to the next, with a half-dozen stages total and with each one lasting several minutes. Scientists know this because they can detect different brain waves at each stage. Under heavy anesthesia, basic sensory signals show up as short-lived, low-frequency pulses—simple stuff. As the patient starts to surface, brain chatter picks up and higher-frequency waves emerge. Pretty soon you see a cascade of signals that instead of dying out quickly, ping back and forth between distant regions. These signals continue to grow in complexity until the patient wakes up fully and the entire brain is humming.

In addition to hinting at how consciousness works, this research could have practical applications. It might help doctors judge the mental torpor of coma patients and determine whether they're still "there" on some level, even if they can't communicate that. These studies might also help eliminate one of the horrors of modern surgery—the fact that people sometimes wake up in the middle of operations. This "anesthesia awareness" happens only rarely—maybe once every thousand surgeries—but it's pretty awful when it does. Victims can feel surgeons slice open their abdomens, shift their internal organs around, and suck out their blood. And because they've been plied with muscle relaxants, they can't alert anyone to their distress. They simply have to endure it, sometimes for hours.

Most victims of anesthesia awareness recall little about the experience afterward; it remains hazy and unreal. But a handful of people recall everything, including the pain, and they suffer from post-traumatic nightmares of being flayed alive. A few end up killing themselves. Personally, I can't think of any worse torture than anesthesia awareness. (Had Dante known about it, he surely would have slipped it into *The Inferno*.) Understanding how the brain

shifts between different stages of consciousness could help eliminate this horror in the future.

Equally important, this work could help solve one of the greatest
and most ancient mysteries of philosophy, how consciousness arises
within the brain. Thomas Beddoes and Humphry Davy may not
have cured any diseases with gases, but ultimately, they wanted to
understand the human psyche as well. If anesthesia really can limn
the depths of human consciousness, then we'll have more reason
than ever to celebrate these wonder-working chemicals of delight.

Interlude: Le Pétomane

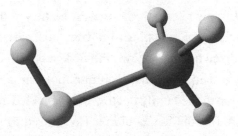

Methanethiol (CH_3SH) — currently 0.000001 part per million of air; you inhale ten billion molecules every time you breathe (unless someone around you has gas)

Dimethyl sulfide (C_2H_6S) — currently 0.00001 parts per million of air; you inhale a hundred billion molecules every time you breathe (unless someone around you has gas)

The past few chapters have examined how gases affect our biology, everything from simple breathing to deep brain function. But you might have noticed a glaring omission so far: we've hardly talked at all about flatulence. Now, don't act like you didn't see this coming. This is a book about gases in all their variety, and there's no gas we think about more often than the gas we pass. So we might as well loosen up, admit that we're all curious, and have some fun with the topic. Joseph Pujol certainly did...

Once upon a time in the South of France, a teenager named Joseph Pujol was playing in the surf at the beach. As he bent over and prepared to dive beneath the waves, he took a huge breath — and felt an icicle of cold stabbing him from the inside. With growing horror, he realized he'd somehow "inhaled" a buttful of water.

The water gushed out of his rectum a moment later, and he felt fine. Still, he ran to the family doctor, who chuckled and told him to forget about it. But the boy couldn't. He refused to go swimming anymore, and he never breathed a word of the incident until his early twenties, when he entered the French army. There, in one of those crude gross-out sessions that seem to break out whenever young men gather, Pujol described what had happened to him that day on the beach. His comrades thought the story hilarious and dared him to try again. Curiosity overcoming fear, he tromped down to the beach on his next furlough and found that he could give himself a cold salty enema at will.

All of this would have been little more than an odd anecdote, except that one fateful day — no one knows how — Pujol discovered that he could pull the same trick with air. He first bent over double, which made it hard to breathe. He then plugged his nose and mouth and contracted his diaphragm, which expanded the volume of his abdomen. Volume and pressure are intimately related with gases: as one rises, the other falls. So expanding his abdomen necessarily lowered the pressure inside it, creating a partial vacuum. Normally when the diaphragm does this, air rushes in and fills our lungs. But because Pujol had bent over and plugged his mouth, air rushed in through the out door instead. Best of all, after inhaling the air, Pujol ripped loose the most epic fart of his life. He was thrilled — and ran off to show his fellow soldiers.

Pujol spent the next several years perfecting this "skill," until he could fart without interruption for ten to fifteen seconds. He found he could vary the pitch and volume of farts as well, playing them like musical notes. He'd always been a theatrical boy, constantly singing and dancing, and after he honed his repertoire in the barracks, he grew a mustache and hit the road with his act in the mid-1880s. He dubbed himself Le Pétomane — the Fartomaniac.

He finally worked up the courage to audition for the famous Moulin Rouge nightclub in Paris in 1892. The job interview consisted of him dropping his trousers, cleansing his "instrument" by sucking up water (he normally gave himself five enemas per day), then serenading the owner. The owner, flabbergasted, hired him immediately.

Audiences didn't know what to make of Le Pétomane at first, but within two years he'd become the highest-paid performer in France, earning 20,000 francs for some shows, more than double what the legendary actress Sarah Bernhardt did. When the curtain opened, he appeared onstage in a black satin tuxedo with white gloves and a red cape. (He wore the tuxedo partly for the incongruity and partly to hide the strain of forcing himself to fart over and over.) After the first laughs died down, he'd do impressions. A short, high toot for a little lass. A meaty blat for a mother-in-law. A bride on her wedding night (a shy peep), and then a few months into the marriage (a thunderclap). He imitated roosters, owls, ducks, bees, toads, pigs, and a dog whose tail got caught in a door. He brought the house down when he

Le Pétomane, the "fartomaniac," could sing and do impressions with his anus.

played a flute backward. Near the end (ahem), he stepped offstage and reemerged with a tube inserted into his anus, like a tail. The other end had a lit cigarette wedged into it, and he proceeded to blow smoke rings out of both ends at once. For the finale, he gave a rousing rendition of "La Marseillaise," then puffed out a candle from several feet.

Women in the audience, especially those wearing corsets, often laughed so hard they passed out. One man had a heart attack. (The Moulin Rouge took advantage of this by stationing nurses around the theater and by posting warnings about how dangerous the show was—which of course only made people more eager to attend.)

And lest you feel guilty for giggling over toilet humor here—or are too stuffy to see what's so funny—know that Le Pétomane hobnobbed with Renoir and Matisse, and that Ravel adored him. Legend has it that Freud kept a picture of Le Pétomane on his wall, and drew on him when developing his theory of anal fixation. Even the king of Belgium came to see the "fartiste" once, albeit incognito.

So what was going on back there with Le Pétomane? To start with, let's examine what farts arc. Partly air. Every time you swallow food or water you ingest a few milliliters of air. Most of this gets burped back up. But some fraction, especially if you're lying down, seeps into the stomach and intestines, where it starts to migrate south.

Some 75 percent of every fart is produced "in house" by bacteria in the gut, through a process called fermentation. We lay folk normally associate fermentation with beer, but there's actually a lot more to it: fermentation covers a wide variety of cases in which carbohydrates get digested and broken down into smaller metabolites. In this case, fart-fermenting bacteria swallow and break down chains of carbohydrates into carbon dioxide, hydrogen, and methane; the microbes then belch them back up, filling our guts with gas. We call certain foods gassy because they contain carbs (the lactose in milk, for example, and the raffinose in cabbage and broccoli) that don't break down much in the stomach and therefore provide a feast for these microbes in our intestines.

The average adult farts around three pints of gas each day, in roughly twenty parcels. Those numbers can vary widely, though. Le Pétomane inhaled almost six pints with each butt breath, and one man in the medical literature cut the cheese ten dozen times each day. Such vast quantities of gas can rupture the intestines if they get trapped and can't find an outlet. In one gruesome case, surgeons cauterizing a man's colon ended up igniting a pocket of gas, blowing a six-inch hole in his abdomen.

Surprisingly, more than 99 percent of the gas in farts has no smell. Even methane, despite its bad reputation, has no odor. Rather, most of the pungency of gas comes from a few trace components: hydrogen sulfide (H_2S), which reeks of rotten eggs; methanethiol (CH_3SH), which stinks of rotting vegetables; and dimethylsulfide (CH_3SCH_3), which smells cloyingly, sickeningly sweet. They're collectively known as volatile sulfur compounds, and they emerge, once again, from the oversated bellies of bacteria. These same gases also contribute to the stink of morning breath. (Try not to think too hard about that.)

So that's how farts work. But it might surprise you to learn that none of that has anything to do with Le Pétomane. Remember, he wasn't scarfing broccoli or chugging raw milk to fill his guts with gas. He was just inhaling good old everyday air and pushing it right back out. As a result, the Fartomaniac's farts, at least onstage, didn't smell. In fact, he eschewed scatological humor in his shows, thinking it vulgar. Rather, he considered himself something between a singer and an impressions guy, like a vaudeville Man of a Thousand Voices.

Which brings up a serious, if startling, question. Both ends of the body have tubes that allow air to pass through. So why don't we "speak" through our rear ends?* There's no a priori reason why not. The anus lacks vocal cords, of course, but those are nothing special, just flaps to help keep food and water out of our airways; the anal sphincter can buzz like a trumpeter's mouth anyway, which is almost as good. A bigger problem is that our butts lack lips and tongues (thank goodness), the organs that mold emerging air currents into words. But really, we could get by with far less complicated oral equipment for basic communication. Overall, then, evolution easily could have done a U-turn long ago and developed fancier folds around our anuses for rectal talking. A few species of herring reportedly do communicate by farting.

After a few glorious years at the Moulin Rouge, Le Pétomane got into some legal skirmishes with the owner. First he got caught crooning at a gingerbread stall in the marketplace, trying to attract business for a friend. After this the owner sued him for breach of contract, arguing that Le Pétomane was allowed to fart only inside the club. (Newspapers found this uproarious.) The Fartiste eventually quit and founded his own club, which prompted more legal wrangling, especially when the Moulin Rouge owner found a female fartomaniac to replace him. (Turns out she was a fraud: she had a bellows secreted beneath her petticoat. Le Pétomane had actually stripped naked before several doctors and undergone an examination early in his career when people accused him of the same thing.)

Although no longer the highest-paid act in France, Le Pétomane made a comfortable living at his new club for two decades, until 1914 turned Europe upside down. No one had time for such frivolity anymore, not even within Le Pétomane's family: two of his sons were maimed on the front lines. After the notorious chemical warfare attacks of World War I, gas-based comedy seemed in rather poor taste anyway.

After the war, Le Pétomane settled down and opened a bakery; he apparently made the best bran muffins around. When he finally died, a month after V-E Day in 1945, several doctors asked the family whether they could examine his plumbing to see what allowed him to inhale gases that way. Sadly, we'll never know, since the family refused permission. As one of his sons explained, "There are some things in this life which simply must be treated with reverence."

CHAPTER FIVE

Controlled Chaos

Water (H_2O) — variable, depending on the landscape and weather; you inhale anywhere from a few billion to several quadrillion molecules every time you breathe

In general, a summons to stand in judgment before the Holy Roman Emperor was not an occasion for celebration. But Otto Gericke, the mayor of Magdeburg, Germany, felt his confidence soar as his cart rattled south. After all, he was about to put on perhaps the greatest science experiment in history.

Gericke, a classic gentleman-scientist, was obsessed with the idea of vacuums, enclosed spaces with nothing inside. All most people knew about vacuums then was Aristotle's dictum that nature abhors* and will not tolerate them. But Gericke suspected nature was more open-minded than that, and he set about trying to create a vacuum in the early 1650s. His first attempt involved evacuating water from a barrel, using the local fire brigade's water pump. The barrel was initially full of water and perfectly sealed, so that no air could get in. Pumping out the water should therefore have left only empty space behind. Alas, after a few minutes of pumping, the barrel staves leaked and air rushed in anyway. He next tried evacuating a hollow copper sphere using a similar setup. It held up longer, but halfway through the process the sphere imploded, collapsing with a bang that left his ears ringing.

The violence of the implosion startled Gericke—and set his mind racing. Somehow, mere air pressure—or more precisely, the difference in air pressure between the inside and outside of the sphere—had crumpled it. Was a gas really strong enough to crunch metal? It didn't seem likely. Gases are so soft, after all, so pillowy. But Gericke could see no other answer, and when his mind made that leap, it proved to be a turning point in the human relationship with gases. For perhaps the first time in history, someone realized just how strong gases are, how muscular, how brawny. Conceptually, it was but a short step from there to steam power and the Industrial Revolution.

Before the revolution could begin, however, Gericke had to convince his contemporaries how powerful gases were, and luckily he had the scientific skills to pull this off. In fact, a demo he put together over the next decade began stirring up such wild rumors in central Europe that Emperor Ferdinand III eventually summoned Gericke to court to see for himself.

On the 220-mile trip south, Gericke carried two copper hemispheres; they fit together to form a twenty-two-inch spherical shell. The walls of the hemispheres were thick enough to withstand crumpling this time, and each half had rings welded to it where he could affix a rope. Most important, Gericke had bored a hole into one hemisphere and fitted it with an ingenious air valve, which allowed air to flow through in one direction only.

Gericke arrived at court to find thirty horses and a sizable crowd awaiting him. These were the days of drawing and quartering convicts, and Gericke announced to the crowd that he had a similar ordeal planned for his copper sphere: he believed that Ferdinand's best horses couldn't tear the two halves apart. You could forgive the assembled for laughing: without someone holding them together, the hemispheres fell apart from their own weight. Gericke ignored the naysayers and reached into his cart for the key piece of equipment, a sort of cylinder on a tripod. It had a tube coming off it, which he attached to the one-way valve on the copper sphere. He then had several local blacksmiths—the burliest men he could find—start cranking the machine's levers and pistons. Every few seconds it wheezed. Gericke called the contraption an "air pump."

It worked like this. Inside the air pump's cylinder was a special airtight chamber fitted with a piston that moved up and down. At the start of the process the piston was depressed, meaning the chamber had no air inside. Step one involved a blacksmith hoisting the piston up. Because the chamber and copper sphere were connected via the tube, air inside the sphere could now flow into the chamber. Just for argument's sake, let's say there were 800 molecules of air inside the sphere to start. (Which is waaaaay too low, but it's a nice round number.) After the piston got hoisted up, maybe half that air would flow out. This left 400 molecules in the sphere and 400 in the chamber.

Now came the key step. Gericke closed the one-way valve on the sphere, trapping 400 molecules in each half. He then opened another, separate valve on the chamber, and had the smithy stamp the piston down. This recollapsed the chamber and expelled all 400 molecules. The net result was that Gericke had now pumped out half the original air in the sphere and expelled it to the outside world.

Equally important, Gericke was now back at the starting point, with the piston depressed. As a result he could reopen the valve on the sphere and repeat the process. This time 200 molecules (half the remaining 400) would flow into the chamber, with the other 200 remaining in the sphere. By closing the one-way valve a second time he could once again trap those 200 molecules in the chamber and expel them. Next round, he'd expel an additional 100 molecules, then 50, and so on. It got harder each round to hoist the piston up— hence the stout blacksmiths—but each cycle of the pump removed half the air from the sphere.

As more and more air disappeared from inside, the copper sphere started to feel a serious squeeze from the outside. That's because gazillions of air molecules were pinging its surface every second. Each molecule was minuscule, of course, but collectively they added up to thousands of pounds of force (really). Normally air from inside the sphere would balance this pressure by pushing outward. But as the blacksmiths evacuated the inside, a pressure imbalance arose, and the air on the outside began squeezing the hemispheres together tighter and tighter: given the sphere's size, there would have been 5,600 pounds of net force at perfect vacuum. Gericke couldn't have known all these details, and it's not clear how close he got to a perfect vacuum. But after watching that first copper shell crumple, he knew that air was pretty brawny. Even brawnier, he was gambling, than thirty horses.

After the blacksmiths had exhausted the air (and themselves), Gericke detached the copper sphere from the pump, wound a rope through the rings on each side, and secured it to a team of horses. The crowd hushed. Perhaps some maiden raised a silk handkerchief and dropped it. When the tug-of-war began, the ropes snapped taut and the sphere shuddered. The horses snorted and dug in their hooves; veins bulged on their necks. But the sphere held—the horses could not tear it apart. Afterward, Gericke picked up the sphere and flicked

The Magdeburg drawing-and-quartering experiment, developed by Otto von Guericke (inset). (Photo courtesy Wellcome Trust)

a secondary valve open with his finger. *Hissss.* Air rushed in, and a second later the hemispheres fell apart in his hands; like the sword in the stone, only the chosen one could perform the feat. The stunt so impressed the emperor that he soon elevated plain Otto Gericke to Otto von Guericke, official German royalty.

In later years von Guericke and his acolytes came up with several other dramatic experiments involving vacuums and air pressure. They showed that bells in evacuated glass jars made no noise when rung, proving that you need air to transmit sound. Similarly, they found that butter exposed to red-hot irons inside a vacuum would not melt, proving that vacuums cannot transmit heat convectively. They also repeated

the hemisphere stunt at other sites, spreading far and wide von Guericke's discovery about the strength of air. And it's this last discovery
that would have the most profound impact on the world at large. Our
planet's normal, ambient air pressure of 14.7 pounds per square inch
might not sound impressive, but that works out to one ton of force per
square foot. It's not just copper hemispheres that feel this, either. For
an average adult, twenty tons of force are pressing inward on your
body at all times. The reason you don't notice this crushing burden is
that there's another twenty tons of pressure pushing back from inside
you. But even when you know the forces balance here, it all still seems
precarious. I mean, in theory a piece of aluminum foil, if perfectly balanced between the blasts from two fire hoses, would survive intact.
But who would risk it? Our skin and organs face the same predicament
vis-à-vis air, suspended inside and out between two torrential forces.

Luckily, our scientific forefathers didn't tremble in fear of such
might. They absorbed the lesson of von Guericke—that gases are
shockingly strong—and raced ahead with new ideas. Some of the projects they took on were practical, like steam engines. Some shaded friv-
olous, like hot-air balloons. Some, like explosives, chastised us with
their deadly force. But all relied on the raw physical power of gases.

The Industrial Revolution began with a broken toy. In the late 1750s
Glasgow University hired a moody craftsman named James Watt to
make and maintain equipment for class demonstrations. One task
involved fixing a small Newcomen steam engine, a toy version of
the engine that mines used to pump water. It stood two feet tall, was
made of brass, and had never worked right. The professor in charge
just wanted Watt to fix the stupid thing, but the more Watt studied
the tiny steam engine, the more he saw ways to *improve* it. This
would grow into a lifelong obsession.

Human beings have long used water to drive machinery,* but steam per se didn't find much use in industry until an English chap named Thomas Savery built a device in 1696 for miners in Cornwall. Although Cornwall had tons upon tons of tin and other mineral wealth, the mine shafts there inevitably filled up with water after a few feet of digging. Bailing this water out involved hitching dozens of horses or oxen to a giant wheel and having them winch buckets up and down—slow, tedious, expensive work. So Savery invented a machine to do the chore. It consisted of two parts, a von Guericke–style vacuum pump to raise water initially, then a second pump that used compressed steam to push the water higher. When peddling it to customers, Savery called the device the Miner's Friend, as if it were a pet dog. It was more like a pet dragon: it stood dozens of feet tall and had flames burning in its belly to make the steam. In fact, in his patent application, Savery gave his machine a dramatic, almost mythical name: "an engine for raising water by fire."

Though better than oxen, the Miner's Friend had its shortcomings. The first involved Savery's vacuum pump, which couldn't raise water past thirty-four feet. The reason for this limit is subtle, and it involves a common misunderstanding about how vacuum pumps work. Most people back then assumed that vacuum pumps raised water by somehow sucking fluids upward. No. Technically, vacuums can't lift or suck anything—which makes sense if you think about it. There's literally nothing in a vacuum, so how could it exert any force or do any work? What really happens is this: whenever you evacuate all the air from inside a vacuum chamber, there's suddenly nothing *pushing back* on any fluids trying to enter from the outside. As a result, those fluids can rush into the void unimpeded. Overall, then, vacuum pumps don't pull fluids into themselves; they simply allow outside fluids to push their way in.

So what's doing the "pushing" when you're raising water? Air pressure. Envision the following setup: a mine shaft with a pool of water at the bottom; a vacuum pump on the ground above; and a pipe leading from the pump down into the pool. As soon as you activate the pump, water begins to rise inside the pipe. But that's not because the pump somehow pulls the water upward. Rather, it's because there's air pressure pushing down on the surface of the pool. Now, that probably sounds backward: air pressing *downward* will *raise* water? But it's true. As an analogy, imagine a lump of bread dough sitting on a counter. Put your hands on the dough and push downward. What happens? Some dough gets compressed, but some will come squirting up through the cracks between your fingers. In other words, pushing down in some places raises the dough in others. The same thing happens with water and air pressure. Air pushes down on most of the surface, but that downward thrust will force the water upward into any region of low pressure—like the vacuum inside the pipe.

All these distinctions about vacuums versus air pressure and pushing versus pulling might seem pedantic at first. And when you're talking about how, say, a nine-inch soda straw works, they probably are. (Please don't start pointing out to people that it's air pressure, not their lungs, raising the liquid in their drinks each time they take a sip. Even though it's true.) But at heights of a few dozen feet, such distinctions become vital. That's because at those heights the downward tug of gravity starts to become a factor.

To see this struggle between air pressure and gravity, imagine a mine worker trying to raise some water with one of Savery's vacuum pumps. I'll spare you the calculations, but if savery sucked up a foot of water into his pipe, the force of gravity pulling downward on that water would amount to 0.4 pounds per square inch. The air pressure holding the column up is 14.7 psi, and because 14.7 is greater than 0.4, air pressure wins here. If he sucks up two feet of water into the pipe, the gravitational pressure doubles to 0.8—still less than 14.7. But at some point, as he sucks up more and more water, the downward gravitational pressure will exceed 14.7, and the column will stop rising. And if you do the math, this happens at around 34 feet.

This is the limit Thomas Savery ran into. Indeed, no vacuum pump on Earth,* even a perfect one, can dead-lift water higher than 34 feet; our atmosphere simply lacks the muscle. Savery's vacuum pumps—which weren't even close to perfect—did even worse than that, topping out below 30 feet.

At this point you might be wondering about the other half of the Savery engine, which used a blast of steam to push water upward through a pipe. There is some good news here, since this method has no intrinsic limit. As long as you keep raising the steam pressure, you can theoretically pump water to the moon this way. The problem was, the valves and joints of the 1690s couldn't handle much pressure, and everything crapped out after another 30 feet. Overall, Savery's devices couldn't lift water more than 60 feet. Better than a horse, but not superb.

A blacksmith named Thomas Newcomen finally invented a superior engine in the 1710s. It once again consisted of two halves. One half had a piston inside a chamber, which moved up and down based on fluctuations in steam pressure. To see how it worked, imagine the piston in the up position to start. There's a soft cloud of steam beneath it, helping to support its weight, and as long as the steam remains there, the piston won't move. Just when the piston starts to relax, though, a valve opens up below it and ruins everything. This valve squirts cold water into the cylinder holding the steam. The cold shocks the steam—it's like someone dumping ice down your shirt—and the gas shrivels and condenses into water, which then drains away through a pipe in the floor. (From the outside, this step reportedly sounded like someone snuffling.) More important, when the steam condenses, there's nothing supporting the piston anymore. It begins to plummet, pushed downward by external air pressure. Before it crashes into the bottom of the cylinder, though, another valve opens and saves the day. Steam comes rushing back into the chamber, buoying the piston and allowing it to rise. Crisis averted, the piston can once again relax on that cloud of steam—until that dastardly valve opens again, squirts in more cold water, and starts the cycle over.

Now, technically, the piston side of the engine didn't pump any water; it just provided the power to. The actual pumping took place on the other side. Above the piston lay a gigantic wooden beam that seesawed up and down like an oil derrick. (Miners imported these 40-foot behemoths from the virgin forests of British Columbia or the Baltic.) One end of the beam was chained to the piston, so every time the piston cycled, the seesaw rose and fell. This rising and falling in turn powered the pump, which was chained to the seesaw's far side. To be honest, though, the "pumps" employed here barely deserved the name. Many were little more than buckets that ratcheted the water upward a few feet each cycle. But unlike fancy vacuum pumps, buckets have no intrinsic limit due to air pressure. And however crude the buckets seemed, the business end of the Newcomen engine—the steam pistons—concentrated power so well that miners could hoist water 150 feet, a miraculous height in the early 1700s.

Nevertheless, for all its benefits, few miners counted the Newcomen engine a close personal friend. The engines cost the princely sum of £1,000 to install. Worse, they gobbled up ungodly amounts of coal every single day. Big mines couldn't afford not to use them, but the fuel costs alone left them teetering on the brink of bankruptcy. The engines were less friend than extortionist, but no one could find a way around them.

Which brings us back to James Watt. While mending the broken toy for his university, Watt couldn't help but notice several inefficiencies in the Newcomen design—especially the excess fuel consumption. Today we'd talk about this inefficiency in terms of heat loss and wasted energy, but scientists back then didn't have a clear picture of those concepts. Watt spoke instead of "wasting steam."

His phrasing sounds quaint today, but there was a lot of truth to this. In that era, steam really was a precious commodity.

Lacking scientific training, Watt threw himself into the problem of wasted steam higgledy-piggledy. In the morning he might collect steam from teapots and melt things. That night he might bury himself in books on phlogiston theory or conduct chemistry experiments. He finally made some headway after chatting with Joseph Black, the lively Scottish chemistry professor who'd discovered carbon dioxide. Black also studied phase transitions, the passage of a substance from, say, solid to liquid or liquid to gas. It was this work that finally provided the insight Watt needed.

To see why, imagine a pot of cold water on the stove. Start warming it from 32°F to 33°F, then 33°F to 34°F, step by step all the way up to 212°F. It just so happens that each one-degree jump takes the same amount of energy, roughly a quarter of a (food) calorie per pound of water. But if you try to heat the water past 212°F, something odd happens. Water boils at 212°F, so water at 213°F is steam. And you might think, based on the steady input of heat required before, that jumping from 212°F to 213°F would take the same amount of energy as jumping from 211°F to 212°F. Not even close. There's an enormous gulf between liquid water and steam. It actually takes far less energy—five times less—to raise the temperature of a cup of water from 32°F all the way to 212°F than to turn that water at 212°F into steam. It's like a marathon that ends on a mountain: the first twenty-five miles aren't nearly as hard as the last, because steam absorbs so much energy.

Black called the extra energy inside steam its "latent heat," and Watt latched onto the idea to explain the inefficiency of Newcomen engines. Again, the Newcomen cycle started with the piston up. Then the steam supporting it condensed, thanks to a squirt of cold water into the cylinder. Obviously, for this to work, the cold water had to cool the steam below 212°F. Less obviously, the water had to cool the piston and the cylinder below 212°F as well. Otherwise any water droplets that condensed onto those hot metal surfaces would hiss and turn right back into steam. This meant squirting a lot more cold water into the cylinder than you might think, which wasted time and energy. But what really killed you was the reverse process. After the piston fell, the engine had to raise it up again with more steam. And unfortunately, having just been chilled with the squirt of cold water, the metal surfaces will now shrivel and condense any steam that comes into contact with them. In other words the cool metal surfaces will suck energy out of the steam like a vampire, leaving behind drops of liquid that do no useful work. To overcome this problem, Newcomen engines had to pump even more steam in, which meant boiling more water, which meant burning more coal, which meant wasting more money. We aren't talking pennies on the dollar, either. Watt calculated that Newcomen engines squandered 80 percent of their steam. Four out of every five dollars pissed away. This is why Watt lamented all the "wasted steam" in the design: steam stores so much latent energy that anything that undermines steam efficiency undermines the entire operation.

To fix this problem, Watt had to find a way to avoid repeatedly heating and cooling the cylinder. Before he could do so, however, life intervened. With a wife and young children to feed, he had to quit tinkering with steam engines and take a job doing survey work for canals. These years weren't a total waste. Like mine shafts, canal holes often fill up with water during the digging phase and need draining, so Watt got to spend several years working with Newcomen engines. But this mostly just enraged him—all that steam wasted, day after day after day.

During these years Watt also began corresponding with a group of gentlemen-scientists who lived 250 miles south of Glasgow, in Birmingham. In some ways Birmingham was the Silicon Valley of its day—a tech wonderland—and tourists from across Europe streamed in to gape at the waterwheels and automatic looms. That said, it wasn't quite Utopia. Workers wandered about smelling of oil, with bloodshot eyes and permanent coughs. Some had hair tinged green from copper smelting.

Despite his own working-class background, Watt fell in with Birmingham's elite, especially its famous intellectual club, the Lunar Society. (Joseph Priestley, William Herschel, and Erasmus Darwin also belonged.) Society members met one evening per month for raucous discussions of literature and philosophy, always gathering on the Monday nearest the full moon. Basing your meetings on the phases of the moon seems charming today, if not downright mystical, but the schedule actually had a prosaic explanation. Members needed moonlight to find their way home afterward.

Watt became especially close with one Lunatic, manufacturer Matthew Boulton. Boulton made a handsome living running a factory that made shoe buckles, silverware, thermometers, and the cheap glass beads that explorers used to swindle Indians out of their homelands. Every winter the brook that powered Boulton's factories froze over, and he either had to shut down or rent scores of horses to turn the mill wheels. Summer droughts presented the same dilemma. So Boulton began wooing Watt to move to Birmingham and develop steam engines to power his machinery.

Watt hemmed and hawed over this offer, until a family tragedy — his wife's death in 1773 — persuaded him to uproot and join Boulton's firm. Watt knew of only one way to assuage grief, by working himself to exhaustion; and in much the same way that he would throw himself into studying gaseous medicine when his daughter Jessie died years later, he threw himself into steam work in Birmingham. It was during this fit of activity that Watt built his famous steam engine.

The key idea had actually occurred to Watt a few years prior, in 1765. Again, in a Newcomen engine, a piston rose and fell based on fluctuations in steam pressure, and in the Newcomen design, both steps took place in the same space, inside the cylinder. But maybe they didn't have to. Watt realized one day that maybe he could

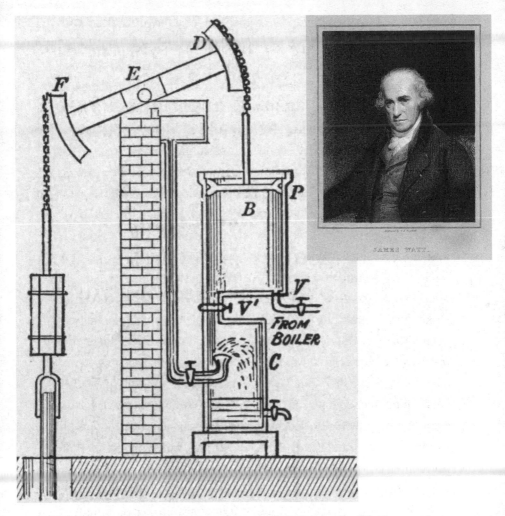

The famous steam engine of James Watt (inset). (Photo courtesy Wellcome Trust)

remove the steam and condense it elsewhere—pipe it into a separate cooling chamber. Removing the steam this way would still allow the piston to fall when the cycle required it. It also allowed him to keep the cylinder piping hot at all times and avoid condensing any steam prematurely.

This vision of a "separate condenser" came to Watt on a Sunday, a

Sabbath, and it killed him to sit on the idea for a day rather than rush into his workshop. Little did he know that it would take him several years to get the separate condenser working properly, and then several years more for him and Boulton to produce a full-fledged steam engine. These delays were mostly Watt's own fault, because his obsessive mind couldn't help but tinker with other parts as well. He flipped the piston upside down and inserted a second steam chamber below it, to push the piston from both sides. He added an air pump to the separate condenser, to suck steam out faster. He fitted the piston with a governor that automatically adjusted its speed, preventing it from running too fast. Cogs within cogs, pipes within pipes: there's no complication Watt wouldn't introduce, provided it wrung out a little more efficiency.

He finally threw together a working steam engine in the mid-1770s — and if you'll forgive a bit of editorializing, the end result strikes me as pretty ugly. I'm a sucker for mechanical simplicity, a quality that von Guericke's air pump and even the Newcomen engine had. Watt's engine lacked grace in comparison — it was a bunch of crap cobbled together. But I can't argue with the results. However much clutter the separate condenser added — extra pipes, extra valves, extra pumps — there's so much latent energy in steam that it was all worth it. To be clear, Watt engines couldn't lift water any higher than Newcomen engines could; but they did that work on one-quarter of the coal, saving mine owners an incredible amount on day-to-day costs. (Imagine your car suddenly running four times farther on a tank of gas.) Watt didn't invent steam power; but if he hadn't made it so economical, the Industrial Revolution likely would have sputtered.

Watt and Boulton sold their first steam engine in 1775, and although it took them several months to assemble each one, their design came to dominate the mining market. They eventually expanded into other markets as well, especially factories. This transition wasn't as straightforward as it might sound, since factories often needed (for historical reasons) engines that provided rotary power, not the up-and-down power that pistons did. After a few years of tinkering, however, Watt came up with several clever designs to meet their demands.

The expansion into new markets got Watt thinking about steam engines in more grandiose terms as well. To most people, the engines were just tools built to accomplish a specific task—pump water, drive a lathe, whatever. Watt envisioned the engines more as *universal* sources of energy—machines capable of powering any mechanical process. As an (anachronistic) analogy, most people saw steam engines as something like calculators: proficient at one task but useless beyond that. Watt dreamed of building the steam equivalent of computers, machines versatile enough to work in any industry.

With these new ideas, Watt found he needed a new vocabulary as well. Factory owners knew of course that Watt could save them money, but they were a conservative lot and they wanted specifics. Sure, a miner could pump 150 feet of water now—but how would that translate to lumber mills and shoe-buckle factories? How many extra yards of lace or bags of flour would they get for their investment?

Rather than calculate every factory's case separately, Watt invented a universal standard of comparison, the horsepower. He defined this rather literally by watching several horses push a mill wheel around and then calculating how far they moved the weight in a certain amount of time (550 foot-pounds per second, he found). This unit was shrewd in several ways. By invoking horses, Watt slyly reminded factory owners what they could give up—all the oats and broken legs and vet bills. Customers also understood the unit intuitively. If ten horses had run their mill wheel before, well, they needed a ten-horsepower engine.

Scientifically, the idea proved prescient as well. Over the next century chemistry and physics would be dominated by thermodynamics, the study of heat and energy. Energy is a vast topic in science, popping up in all sorts of different contexts, and scientists needed a standard unit of comparison to understand how quickly different processes absorbed and released energy. The horsepower fit the bill perfectly. Little did those scientists know that the whole idea started as a marketing scheme by James Watt.

(As thermodynamics branched out into new phenomena, however, like light and magnetic fields, the absurdity of the name "horsepower" became obvious — as if you could still hitch old Bessie to the apparatus. In 1882 physicists finally voted to establish a new universal unit of power, which applies just as readily to lightbulbs and refrigerators as to engines for raising water by fire. They called it the watt.)

Alas, during Watt's lifetime steam never became the universal source of energy that he envisioned. This was partly because the inventor wasted an increasing amount of energy in court battling patent lawsuits. (Watt called patent infringers "imps of Satan," and his legal crusades landed several in prison.) Truth be told, though, Watt himself deserves most of the blame for the stunted growth of steam. Several contemporaries hit upon ways to shrink Watt's huge engines and, say, fit them onto ships with paddlewheels. Rather than embrace these ideas, Watt flung every legal obstacle he could in front of them. A clever engineer at Boulton's factory named William Murdoch built a steam-powered model train in 1784; Watt and Boulton told him to knock it off. Lunar Society member Erasmus Darwin, a prominent physician who put in thousands of miles on England's bumpy roads each year (and had the sore rump to show for it), sketched out a steam-powered automobile. Watt eventually patented the idea — in large part to block others from building one.

Not even Watt standing athwart history and yelling Stop! could stunt the development of steam power forever, though. Steam was simply too dynamic, too vigorous for engineers to ignore. And to give Watt his due, historians have noted that his basic engine design survived unchanged for three-quarters of a century after he died. Think how much computers and other technologies have changed in the past quarter century alone, and you can appreciate how incredible that is.

Watt's steam engines proved so powerful, in fact, that they uprooted English society. Factories were no longer tethered to river power and could move into cities, dragging millions of workers with them. (In the mid-1800s England became the first country in history to have more people living in urban areas than rural ones.) Women and children entered the workforce as well, and a new stratum of society, the middle class, emerged. Rather than the bloody revolution of France, Great Britain rebuilt its society via an industrial one, and James Watt was no small reason why.

————<o>————

Although steam eventually became an all-purpose source of power in the nineteenth century, other gases still trumped steam in certain applications. Take explosives.

For centuries humankind knew of only one explosive, gunpowder, a mix of carbon, sulfur, and nitrates. When burned, those three ingredients react as follows: $3C_{(solid)} + S_{(solid)} + 2KNO_{3(solid)} \rightarrow N_{2(gas)} + 3CO_{2(gas)} + K_2S_{(solid)}$. Don't sweat the details there, but notice that you start with six molecules of solids on the left, and end up with five molecules on the right—four of them gases. It's those gases that propel bullets and shrapnel at high speeds.

Still, getting just four gas molecules from six molecules of starting material isn't that impressive, and an Italian chemist named Ascanio Sobrero finally discovered an alternative to gunpowder in 1846. Like most good chemists then, Sobrero had a little cowboy in him, and he loved working with hazardous materials. In one experiment, he started plopping substances like rubber and lactose into acids, just to see what happened. One chemical he tried was glycerin, a type of sugar, which he dripped into a bath of nitric and sulfuric acids. A pale yellow liquid emerged, which Sobrero compared to olive oil. Because the overall reaction involved fusing nitro groups ($-NO_2$) onto the glycerin, he called the substance nitroglycerin ($C_3H_5O_9N_3$).

As a test, Sobrero put a single drop of nitroglycerin in a sealed test tube and held it over a flame. A moment later, he was picking shards of glass out of his face and hands. We can see why the tube exploded on him by examining the chemical reaction: $4C_3H_5O_9N_{3(liquid)} \rightarrow 6N_{2(gas)} + 10H_2O_{(gas)} + 12CO_{(gas)} + 7O_{2(gas)}$. The key point is that four molecules of nitro yield a whopping thirty-five molecules of gas, a dynamite return on investment. Even better, nitro releases those gases almost instantly,* within a millionth of a second. (To put that in perspective, gunpowder explodes in a few thousandths of a second. Which means that if you stretched out the gunpowder explosion to a full hour, the nitro explosion would finish in four seconds.) As scientists now know, that speed is what makes explosives so deadly. Ounce for ounce, gasoline, coal, and even butter(!) store more energy in their chemical bonds than most explosives. Explosives simply release their energy far more quickly. Explosives also have a means—gases—to hurl that energy outward, and it's the gases that do most of the damage.

Sobrero tried to sell his sugar bomb to the Italian government, but it demurred: in tests, nitroglycerin proved a touch *too good* at destroying things. Even then, nitro might have found an application if not for another problem—its fickleness. Sometimes heat set it off, as Sobrero found. Other times, though, it simply burned without exploding. Still other times, it would go off with no heat at all: a hard jolt might suffice. Or it would combust almost spontaneously, like Dickens's Mr. Krook. This unpredictability made nitro too dangerous for routine use—and all the more alluring for chemical cowboys. They kept eying nitro, circling around it, looking for a way to tame it. And although it took twenty years, a lonely, unhappy Swede named Alfred Nobel finally did.

Nobel's father, Immanuel, manufactured cannonballs, mortars, torpedoes, and other instruments of war, and the family often suffered cruelly during times of peace. Thanks to a factory fire, they went bankrupt the year Nobel was born, in 1833. Immanuel finally persuaded the Russian military to let him build naval mines for them, and he uprooted his wife and four children from Stockholm to St. Petersburg in 1842. It was likely a rough journey for Alfred, a dyspeptic youth with a hacking cough and weak heart. Unfit for a vigorous life, he'd focused on academics and had excelled in several subjects, eventually becoming fluent in German, English, French, Italian, and Russian. Given this facility with languages, he yearned to write, but his father pushed him toward science instead, and he took up chemistry dutifully, if halfheartedly. That all changed one afternoon when his tutor in St. Petersburg arranged a demonstration. He smeared a drop of what looked like olive oil on an anvil and told everyone to step back. The setup couldn't have looked that impressive, might even have elicited a few sniggers from the boys. Then the tutor smacked the anvil with a hammer.

The bang and flash bowled Nobel over, figuratively if not literally. All that power in one drop! The substance fascinated him, and when his father once again went bankrupt after the Crimean War, Alfred convinced most of the family to throw in with him and focus on nitroglycerin.

Insane as it sounds, Nobel decided that the way to make nitroglycerin safer was by combining it with gunpowder. Again, people feared nitro because it seemed to explode at random. Nobel reasoned that nitro just needed a more reliable trigger, and there was no more reliable trigger than good old gunpowder. So he designed a prototype bomb with three components: a small glass vial of nitro, a sealed canister of gunpowder to slip the nitro into, and a fuse. The hope was that the fuse would first ignite the gunpowder, liberating hot gas; the hot gunpowder gas would then slam into the nitroglycerin and detonate it—history's first two-stage explosive. Nobel invited his brothers to watch him test the device in a drainage ditch near their St. Petersburg factory. He lit the fuse, plopped the bomb into the water, and ran. *Kaboom!* Sewage rained down on them, splashing their clothes. Nobel was thrilled.

Over the next year Team Nobel returned to Stockholm, and when Alfred finally turned back to nitro research, he had a new idea for a bigger bang: he would invert his initial setup and put a vial of gunpowder inside a hefty canister of nitro. This time he invited his father and his brothers Robert and Oscar-Emil to watch; and, feeling braver, he decided to detonate this device not in water but in plain air.

On the big afternoon Nobel once again lit the fuse and ran. Three, two, one, aaaand—nothing. The bomb fizzled. The gunpowder went *pffft* and disappeared in a plume of smoke. The nitro puddled harmlessly onto the dirt. Everyone stared for a second, dumbstruck. Then Robert and Immanuel burst out laughing; they doubled over, tearing up. *I guess this setup bombed, eh, Alfred? Hahahaha.* Nobel, meanwhile, fumed. He stomped away and locked himself into his lab to figure out what had gone wrong. Why had the nitro-gunpowder mix exploded underwater but not in plain air? It couldn't be caprice: there had to be a scientific explanation.

He eventually realized that the problem revolved around pressure. In exploding the first bomb underwater, he had inadvertently confined the hot gunpowder gases to a small space, since the water hindered their escape. As a result, the gases had time to slap the nitroglycerin and detonate it. Air, though, couldn't provide the same straitjacket of pressure. In this case the gunpowder gases rocketed away, leaving the nitro unmolested. Nobel realized he therefore needed a way to confine the gunpowder. After a year of tinkering, he came up with what's called a blasting cap, a hollow wooden plug that slowed the escape of gunpowder fumes just long enough to trigger the nitroglycerin. Here was the breakthrough he'd longed for.

Alfred Nobel, dynamite maker and benefactor of the Nobel prizes.

Nobel began peddling both blasting caps and Blasting Oil (his trade name for nitro) in 1864. He'd just secured his first big orders that September, in fact, to help build the Suez Canal, when 250 pounds of nitro went off in a storage shed near his lab. The explosion destroyed several buildings and killed five people, including his twenty-year-old brother, Oscar-Emil, the one who hadn't laughed at him. Nobel had been miles distant at the time, and no one ever determined what set the nitro off, but when the police found out he'd been making nitro illegally inside city limits, they threatened to charge him with murder.

He tried to assure the police that nitroglycerin was safe if handled properly, but the smoldering rubble of his laboratory belied this. A string of similar accidents, at home and abroad, hardened public opinion against him even more. Honestly, some of the victims here sound like candidates for the Darwin Awards: rubes who polished their boots with nitro or greased their wagon wheels with it. The Welsh miner who was playing soccer with a tin of nitro almost deserved it. Other accidents couldn't be dismissed so easily. In 1866 a leaky crate of Nobel Blasting Oil arrived in San Francisco, and a few warehouse hands tried to pry the lid off with a crowbar. They succeeded only in cratering out the block. A severed human arm smacked a third-story window down the street. Several buildings over, rescue workers found an intact human brain lying on the floor.

As these accidents piled up—in New York, in Panama, in Sydney, in Hamburg—Nobel became a veritable public enemy. No one would sell him lab space anymore, so he had to convert a barge into a floating chemical laboratory and work on the water for a few years. The bangs and fumes wafting in from the harbor always gave him away, though, and he'd have to pull anchor and make a run for a new port, fleeing like a fugitive. Once or twice mobs even attacked the "Nobel death ship."

In 1867 Nobel finally hit upon a solution to the problem—a way to take the edge off nitroglycerin without neutering it. Chemically, you can think of nitro as a bundle of gases weakly bound together in liquid form. But the liquid lacked the strength to confine these feral molecules. So Nobel decided to reinforce the liquid by mixing in a solid, kieselguhr, a soft white clay. (It forms underwater when sea critters called diatoms decay.) Various legends, which Nobel always denied, say that he discovered the mixture accidentally one day, after spilling some nitro onto a pile of kieselguhr like a klutz. Whatever the process, Nobel liked the way he could sculpt the resulting putty into sticks. The clay turned out to weaken the nitro, but it also prevented an accidental jolt or shock from blowing the mixture up. And even in this weaker state, it was five times more powerful than gunpowder. He named the substance Dynamite after the Greek word for power.

Dynamite became a big hit—as well as the most controversial commodity in the world, the napalm or Agent Orange of its day. On the one hand mining and construction firms clamored for it: it was perfect for blasting out mountain passes or blowing stubborn reefs out of harbors. London got the world's first subway thanks to it. On the other hand plenty of accidents still occurred, and the potency of Dynamite ensured that rather than one or two workers, a dozen might die at once. One especially gruesome project, a tunnel in Switzerland, cost twenty-five lives per mile. Even worse, Dynamite seemed destined to make wars more deadly in the future. Not as a replacement for gunpowder: nitro was actually too powerful for cannons and guns, and splintered their barrels. But it made for awful land mines and bombs. As a result, Nobel became an international pariah, an easy target for newspaper editorials and politicians in high dudgeon. Here was a man, they thundered, minting his fortune from death.

Nobel managed to make himself even more unpopular through patent litigation. Like James Watt he defended his intellectual property fiercely, and if anything, he had a worse time of it. For one thing Nobel had far more to defend: he received a staggering 355 patents during his lifetime, on both chemicals and blasting equipment; you could almost make a page-a-day calendar of them. Second, the nature of chemistry made the litigation messier. As soon as Nobel secured a patent on some molecule, someone would come along and tweak it— tack a few extra atoms on, twist some side branch in a different direction. The molecule functioned the same basic way, but legally it fell outside the patent's scope. Several companies also made Dynamite knockoffs by mixing nitroglycerin with other clays and calling the result Hercules Powder or Rendrock. With each passing year Nobel spent less time on research and more time suing competitors.

Despite the lawsuits and nasty press, Nobel's business boomed. It turns out that people generally cared a lot more about convenient train tunnels and canals than dead workers. The man once reduced to working on a boat soon commanded an empire of ninety-three factories in twenty-one countries, and he amassed a fortune equivalent to a quarter billion dollars today. Most telling of all, Dynamite dominated the explosives market to such a degree that it became dynamite, a generic, lowercase noun like thermos and zipper.

Riches, however, didn't make Nobel happy. Always aloof, he never married, and he grew more and more distant from his family in middle age. He occasionally revisited his youthful dream of becoming a writer and sketched out a poem or play, but didn't dare publish them. As the world's disgust for him deepened, he began to despise himself as well — for killing his brother, for war profiteering. When asked to contribute a few lines about himself to a family genealogy, he wrote, "Alfred Nobel — a miserable half-life; should have been smothered by a compassionate doctor when he first made his crying entry into life. Greatest merits: keeps his fingernails clean. Greatest faults: has neither a family, a happy temperament, or a good stomach. Greatest and only desire: not be buried alive." At his gloomiest he talked about opening up a luxury "suicide emporium" where people could shuffle off in peace, in private rooms on plush beds, with classical music playing in the background.

Still, he didn't realize quite how much the world loathed him until 1888, when his brother Ludvig died in Cannes. A few days later a French newspaper, mistakenly thinking that Alfred had died, ran an obituary titled "The Merchant of Death Is Dead." Nobel, then living in Paris, shuddered. All those tunnels and harbors and subways he'd help build, all that mineral wealth he'd helped uncover — *and I'm nothing but a common murderer to them.* Determined to salvage his reputation, he amended his will and set up a prize fund to reward outstanding research in chemistry, physics, and medicine. To make up for his lost boyhood dream, he also endowed a prize in literature. And to atone for a lifetime of peddling death, he endowed the Nobel Peace Prize. (Perhaps we all should be so lucky to read our own obituaries…)

Nobel's health, always shaky, finally failed over the next few years. He'd long suffered from headaches caused by inhaling nitroglycerin fumes, a complaint many of the workers at his factories shared. When metabolized by the body, nitro releases nitric oxide gas (NO). NO causes blood vessels to dilate, which floods the cranium with blood and causes pounding headaches. Nobel apparently never became immune, although most people build up a tolerance. His workers in fact used to smear nitro on their hatbands every Friday afternoon and huff a bit over the weekend so they wouldn't lose their tolerance and end up with a headache come Monday morning.

Besides headaches Nobel also suffered from angina pectoris, a buildup of plaque in the coronary arteries. This prevents oxygen from reaching the heart muscles, and severe chest pain results. Ironically, when Nobel finally sought treatment, his doctor prescribed nitroglycerin. Because nitro causes blood vessels to dilate, when it's injected in small, subheadache doses it can tease open the coronary arteries just enough to relieve the agony of oxygen debt. (Some of Nobel's employees had already figured this out: in contrast to the ones who dreaded Monday-morning headaches, those with chest pain loved coming to work. The fumes were free medicine.) Nobel, however, refused to take nitroglycerin at first. The chemical already dominated his thoughts and business dealings, and deliberately injecting it into his body seemed too much. In the end, though, he gave in, allowing this strange, deadly chemical to penetrate even his heart.

Nobel suffered a stroke on December 10, 1896, and was found slumped dead in a chair. A few days later, his executors read his four-page will to his relatives—who were shocked to find "their" money going to some stupid prize fund. Fortunately for them, Nobel had written the will without consulting any lawyers. (After decades of patent litigation, he mistrusted the entire profession.) This left the will open to legal challenges, and years of petty wrangling followed. A secret paramour even emerged and threatened to splash Nobel's love letters all over the tabloids. We normally think of the Nobel prizes as exalted, as the summa of human achievement. But they were founded in vanity—a dying man attempting to whitewash his reputation—and almost didn't happen because of common greed.

In the end Nobel's executors bought off everyone they needed to. The last obstacle to establishing the prizes involved transferring Nobel's fortune from Paris back to Sweden. EFTs didn't exist then, obviously, and no one would insure the transport of that much cash, so one executor started withdrawing stacks of bills and securities on his own, a few million at a time, and stowing them in briefcases. He then grabbed a loaded revolver and personally escorted them across Europe in carriages and trains. Despite all the hassles, the first Nobel prizes were awarded on December 10, 1901, just five years after Alfred Nobel's death. In the century since, they've become the most prestigious prizes in science, and have all but exploded Nobel's reputation as the merchant of death.

————◄○►————

Steam engines and explosives helped catalyze the Industrial Revolution, allowing us to lift thousands of pounds with ease, or crumble in mere minutes a mountain that had stood since the days of the dinosaurs. But they certainly weren't the only important technologies of the era, or even the only important gas-based technologies. Another key advance involved making high-grade metals, especially steel. Now, steel and gases might not seem to have much in common, but we never could have built the bridges, skyscrapers, and supertankers that dominate our world if not for some key insights into the chemistry of air.

Interlude: Steeling Yourself for Tragedy

Carbon monoxide (CO)—currently 0.1 parts per million of air (higher in urban areas); you inhale roughly one quadrillion molecules every time you breathe

You can debate the relative merits of Chaucer versus Milton, Auden versus Yeats. But when it comes to the worst poet in the English language, William Topaz McGonagall stands beneath all.

Audiences in nineteenth-century Scotland turned out in droves to hate-watch his recitals, and it's hard to say which of his literary offenses delighted them most—the clumsy rhymes and rhythms, the vapid insights ("Like most great men, I was born at a very early period of my existence"), the unwavering faith in his own brilliance (he titled his first book *Poetic Gems*). Perhaps it was his shrill, histrionic delivery, or the fact that he appeared onstage wearing a kilt and brandishing a sword. (In his defense, the sword came in handy for swatting away, midsentence and midair, the occasional fish or

William McGonagall, probably the worst poet who ever lived.

rotten apple hurled his way.) But McGonagall's greatest skill was probably for taking genuinely sad moments and deflating them: no one could transform pathos into bathos faster. Take his masterpiece, "The Tay Bridge Disaster." Please.

On Sunday evening, December 28, 1879, a northbound train began crossing the longest bridge in the world, the two-mile span over the Tay estuary in eastern Scotland. Shakespeare set *Macbeth** near here, but the landscape didn't quite inspire McGonagall to the same heights. "When the train left Edinburgh / The passengers' hearts were light and felt no sorrow," he noted, deftly setting the scene. All was not well, however, for "the rain came pouring down / And the dark clouds seem'd to frown." Sure enough, while the train was crossing "the silvr'y Tay," "the central girders with a crash gave way," and the bridge collapsed. All aboard died, and McGonagall bemoaned this "last Sabbath day of 1879, / Which will be remember'd for a very long time." Thanks to McGonagall the disaster has been remembered, although not for the most solemn reasons.

Like most of his contemporaries McGonagall blamed the "wind [as] it blew with all its might" for toppling the bridge. But while the wind did blow mightily that last Sabbath day of 1879, modern chemistry can point to another, contributing factor: carbon monoxide. Or really, the *lack* of carbon monoxide during the production of the bridge's iron towers.

—◦—

We think about iron as a burly metal, but pure iron is actually weak. On a molecular level its atoms form nice smooth layers that look pretty but that tend to slide past one another if you put any pressure on them. This made pure iron—also known as wrought iron— quite malleable and therefore ideal for artwork or engineering tasks that require flexibility. It also made wrought iron useless for bearing heavy loads.

For engineers in the 1800s, the alternative to wrought iron was cast iron. Cast iron contains impurities, mostly carbon atoms, that disrupt the smooth molecular layers and prevent them from sliding around. This made cast iron quite strong, perfect for supporting bridges and large buildings. Unfortunately the carbon also makes cast iron inflexible and brittle. Like ceramics, cast iron is strong up to a point; but past that point it can't bend and will snap under too much strain.

What engineers really needed for large-scale building projects was steel, an in-between metal with around 1 to 2 percent carbon—enough to add strength but not enough to leave the metal brittle. Unfortunately, making steel back then was a gigantic pain in the catookus, since it required a ridiculous amount of processing.

The processing started with iron ore. Ever since the Oxygen Catastrophe a few billion years ago, most of the iron on Earth has been trapped inside ores such as hematite (Fe_2O_3) and magnetite (Fe_3O_4). Liberating the iron requires smelting the ore—that is, heating it with coke (a carbon-rich solid derived from coal) and mixing in a little air. Air contains oxygen gas, which reacts with the carbon in the coke to form carbon monoxide (CO).* Carbon monoxide then wriggles its way into the ore and Pac-Man-gobbles the oxygen atoms there to form CO_2. This process also introduces carbon from the coke into the iron. The initial product of iron smelting is therefore carbon-rich cast iron.

Making wrought iron required additional steps. Workers first melted the cast iron down, an expensive, fuel-intensive process. They then had to dump more iron ore into the batch. Eventually the carbon in the melted cast iron and the oxygen in the melted ore reacted to form more carbon monoxide, which then bubbled away and left pure wrought iron behind. But because this reaction involved mixing liquids and not gases, it took several days. In the meantime someone had to stand there, stirring the whole mess by hand.

At this point most smelters gave up. Very few took the next step of making steel, which required still more processing. First, those dedicated ironmongers mixed more coke into the wrought iron, to add some carbon back in. They then had to cook the whole batch for several weeks. Given all the hassles here, smelters generally produced steel in small, artisan batches only, for tools or blades. No one dreamed of making whole buildings with steel. For that, engineers were stuck with iron.

Henry Bessemer, steel magnate. (Photo courtesy Wellcome Trust)

Such was the art of steelmaking when Englishman Henry Besse-mer appeared on the scene. Bessemer looked like a U.S. Civil War general with big bushy muttonchops, and over his lifetime he accu-mulated 117 patents — on microscopes, velvet, varnish, sugarmak-ing, you name it. In one early triumph, in the 1850s, he invented an elongated projectile shell with brilliant aerodynamics. He thought it might win the Crimean War for the British, but alas, the shell tended to crack the weak cast-iron cannons of the day. An irked but curious Bessemer decided to investigate how iron and steel got made.

The story of Bessemer's discoveries in this field is long and convoluted, and there's no room to get into it all here. (It also involved several other chemists and engineers, most of whom he stingily refused to credit in later years.) Suffice it to say that through a series of happy accidents and shrewd deductions, Bessemer figured out two shortcuts to making steel.

He'd start by melting down cast iron, same as most smelters. He then added oxygen to the mix, to strip out the carbon. But instead of using iron ore to supply the oxygen atoms, like everyone else, Bessemer used blasts of air—a cheaper, faster substitute. The next shortcut was even more important. Rather than mix in lots and lots of oxygen gas and strip all the carbon out of his molten cast iron, Bessemer decided to stop the air flow partway through. As a result, instead of carbon-free wrought iron, he was left with somewhat-carbon-infused steel. In other words, Bessemer could make steel directly, without all the extra steps and expensive material.

He'd first investigated this process by bubbling air into molten cast iron with a long blowpipe. When this worked, he arranged for a larger test at a local foundry—seven hundred pounds of molten iron in a three-foot-wide cauldron. Rather than rely on his own lungs, this time he had several steam engines blast compressed air through the mixture. The workers at the foundry gave Bessemer pitying looks when he explained that he wanted to make steel with puffs of air. And indeed, nothing happened for ten long minutes that afternoon. All of a sudden, he later recalled, "a succession of mild explosions" rocked the room. White flames erupted from the cauldron and molten iron whooshed out in "a veritable volcano," threatening to set the ceiling on fire.

After waiting out the pyrotechnics, Bessemer peered into the cauldron. Because of the sparks, he hadn't been able to shut down the blasts of air in time, and the batch was pure wrought iron. He grinned anyway: here was proof his process worked. All he had to do now was figure out exactly when to cut the airflow off, and he'd have steel.

At this point things moved quickly for Bessemer. He went on a patent binge over the next few years, and the foundry he set up managed to screw down the cost of steel production from around £40 per ton to £7. Even better, he could make steel in under an hour, rather than weeks. These improvements finally made steel available for large-scale engineering projects—a development that, some historians claim, ended the three-thousand-year-old Iron Age in one stroke, and pushed humankind into the Age of Steel.

Of course, that's a retrospective judgment. At the time, things weren't so rosy, and Bessemer actually had a lot of trouble persuading people to trust his steel. The problem was, each batch of steel varied significantly in quality, since it proved quite tricky to judge when to stop the flow of air. Worse, the excess phosphorus in English iron ore left most batches brittle and prone to fracturing at cold temperatures. (Bessemer, the lucky devil, had run his initial tests on phosphorus-free ore from Wales; otherwise they too would have failed.) Other impurities introduced other structural problems, and each snafu sapped the public's confidence in Bessemer steel a little more. Like Thomas Beddoes with gaseous medicine, colleagues and competitors accused Bessemer of overselling steel, even of perpetrating a fraud.

Over the next decade Bessemer and others labored with all the fervor of James Watt to eliminate these problems, and by the 1870s steel was, objectively, a superior metal compared to cast iron— stronger, lighter, more reliable. But you can't really blame engineers for remaining wary. Steel seemed too good to be true—it seemed impossible that puffs of air could really toughen up a metal so much—and years of problems with steel had corroded their faith anyway. When Bessemer once suggested using steel beams to make train tracks, a railroad executive sputtered, "Do you wish to see me tried for manslaughter?" And the all-important British Trade Board, which oversaw public works, banned steel skeletons for bridges. As a result, when construction began on the Tay Bridge in 1871, engineers had no choice but to use cast iron for the support towers and wrought iron for the cross bracing—and hope that their mutual weaknesses would offset each other.

Not that the engineers thought this arrangement would compromise their design. On the contrary, they hailed the Tay as the biggest, strongest bridge ever built—the *Titanic* of architecture. Unfortunately, several factors conspired to undermine the bridge. First was miscommunication between the foundry that supplied the cast-iron ingots and the construction firm that actually poured and cast the towers. In a scene right out of *Catch-22,* the construction firm requested the "best" iron available for some components. Little did they know that the foundry sold three grades of iron ingots: best, best best, and best best best. So in asking for the "best," the construction firm got the worst. During the pouring and casting, structural bits on several towers snapped off and had to be welded back on; the towers also ended up pitted, as if rotten. Rather than fall behind schedule, the workers who erected the towers filled the holes with a putty made of several not-so-load-bearing materials—rosin, beeswax, iron filings, and of course lampblack, to conceal their subterfuge. (Workers called this putty "Beaumont's egg," a corruption of its French name, *beau montage.*)

After the bridge opened, lazy inspectors compounded the danger by not reporting several cracks that had developed. A historian described one inspector's test thus: "He wetted a piece of paper from his notebook with saliva, pasted it across the crack, and waited for the next train. No tears in the paper. No problem!" To make matters worse, the bridge had been poorly designed in the first place, top-heavy and prone to swaying: even on calm days, it rocked sideways four to six inches whenever trains passed over. Furthermore, the head engineer, Thomas Bouch, had bungled a few calculations and hadn't reinforced the bridge properly for gusts.

As a result of all this, on December 28, 1879, with gale-force winds (sixty-five miles an hour) bearing down on it, the Tay Bridge was reeling like a palm tree in a hurricane. When a small passenger train crossed at six p.m. it nearly slipped off the track, skidding against the guardrails and sending up sheets of sparks. It barely made it across.

The hundred-ton express train an hour later wasn't so lucky. It

The destruction of the Tay Bridge in Scotland.

too smacked the guardrails, igniting an even larger shower of sparks. Right on cue, the wind kicked up at that very moment, slamming the structure at the worst possible time. A bridge made of strong, flexible steel might—might—have held. The stiff, crappy cast iron had no chance. All twelve central towers collapsed— *boom, boom, boom, boom*—opening up a half-mile gap in the span. The train sailed forward into empty air, plunging eighty-eight feet into the water and killing all seventy-five passengers.

The government soon opened an inquiry, and every sordid detail about the cut-rate construction and "treacherous" (their word) cast iron came to light. Needing a scapegoat, investigators sacrificed Bouch, the head engineer. Never mind that he'd just been knighted that summer (along with, coincidentally, Henry Bessemer) or that he lost his son-in-law in the accident. Already sick and feeble, Bouch broke down. He died just months later.

While everyone was distracted, the Board of Trade quietly rescinded its ban on steel bridges. In fact one of the very next bridges it approved, the replacement span over the Tay, used steel for its support towers. It opened in 1887 and still stands today. Naturally, William Topaz McGonagall wrote another poetic gem to commemorate its opening.

The previous chapter's stories about steam and explosives emphasized the power of gases, and steel production seems to reinforce that theme: an infusion of carbon monoxide and oxygen, after all, is what transforms brittle iron ore into strong steel. Really, though, there's a better lesson to draw here, about the elegance of gases.

A much better poet than William McGonagall, e. e. cummings, captured this sentiment perfectly in a different context. In a famous poem cummings marvels at the feelings a lover awakens inside him, "petal by petal…as Spring opens / (touching skilfully, mysteriously) her first rose." He insists that "nobody, not even the rain, has such small hands." It's a lovely way to think about rain—how it trickles into the soil and stirs the buried life there, working in a dimension we can barely fathom. Gases do the very same thing. Whether we're talking about the pas de deux of gases in our lungs or the delicate surgery that CO and O_2 perform in steel, their alchemy seems no less mysterious. Gases have small hands, too.

CHAPTER SIX

Into the Blue

Helium (He)—currently five parts per million in the air; you inhale 70 quadrillion molecules every time you breathe

Argon (Ar)—currently one percent of the air (10,000 parts per million); you inhale a hundred quintillion molecules every time you breathe

Beyond enabling us to make great material progress, new experiments with air touched the human spirit as well. We suddenly understood what this mysterious, magical gas all around us was, giving us deeper insight into how our planet worked. Even more amazing, in the late 1700s gases fulfilled probably the most ancient dream of humankind—allowing our flat-footed, wingless, and decidedly terrestrial species to take to the skies for the very first time.

An early Montgolfier balloon, filled with hot air and smoke.

According to legend, the story of human flight began with lingerie. By the time he turned forty-two, Joseph-Michel Montgolfier had mostly failed in life. Although heir to a papermaking fortune, he'd served time in a debtors' prison in central France, then endured the humiliation of his younger brother taking over his share of the family business. Montgolfier's passion, studying the pneumatic chemistry of Priestley and Lavoisier, further marked him as a misfit in people's eyes. One afternoon in 1782, however, that study of gases paid off most unexpectedly. Watching his wife's laundry dry over a fire, he noticed that her unmentionables kept billowing outward suggestively. Hot dog. He also noticed that they kept rising whenever the fire kicked up. Why? What made them lift? He began to wonder if he could ever build a "sack of air" large enough to hoist himself. In the midst of what amounted to a naughty little daydream, Montgolfier envisioned the world's first balloon.*

Without stopping to think (sometimes a virtue), he built a rectangular box of wood and covered it in silk. It weighed five pounds and stood four feet tall, and when he held it over a small, smoky fire inside his home, it floated to the ceiling. He tried a similar test outdoors and watched it climb seventy feet high. According to most historical accounts, Joseph was a clumsy chemist — but somehow these experiments worked. He was ecstatic: for once he wasn't failing.

At this point Joseph showed the contraption to his younger brother, Jacques-Étienne, and you have to admire Étienne's devotion to his ne'er-do-well sibling. Putting aside any doubts, he helped Joseph build a larger, fifteen-pound box. And — sacré bleu! — when they let it fly, it snapped its tether and drifted for a mile. They chased after it, only to see some sourpuss pedestrian destroy it when it landed. The brothers nevertheless walked away elated. Real flight! Over the next few months they began sketching out plans to make an even larger balloon, one they hoped would make them famous.

This balloon, an experiment through and through, would measure thirty feet across, and for the first time it would be spherical. For the balloon envelope, the Mongolfiers tried buttoning together strips of silk lined with stiff paper. It worked surprisingly well. But their other main experiment, with new lifting gases, proved a struggle. They first tried the steam then causing such a buzz in England.

It only made the paper soggy. They switched to Cavendish's hydrogen gas, only to realize that no one had ever produced more than a few liters at once; they would need several thousand. Worse, as the tiniest molecule in existence, hydrogen easily diffused through their silk envelope and left their balloon flaccid. They finally switched back to hot air in 1783, albeit with a twist. You see, despite his love of chemistry, Joseph actually understood very little about gases. For one thing, he thought that heating air and adding smoke to it magically changed its chemical properties. (In truth smoke isn't a gas at all; it's a suspension of solid particles within air, like murky water.) Moreover he believed that smoke, and smoke alone, created lift: he famously described a balloon as "a cloud in a paper bag," and while that sounds poetic, he probably meant literal clouds of smoke. So for the first public demonstration of their unmanned balloon, in June 1783, the Montgolfiers made the smokiest bonfire possible, fueled by straw, wool, rabbit skins, and old shoes.

Thousands of people crowded around anyway, and watched through choking tears as the balloon envelope rose and took shape. Several stout men held it down with ropes until someone gave the signal and they let go. It sprang up, and soared overhead like a miniature moon.

Although the demo made the Montgolfiers famous, it had the unwelcome side effect of attracting rivals—rivals who knew what they were doing. After a little thought, a second pair of brothers who lived near Paris, the engineers Anne-Jean Robert and Nicolas-Louis Robert, devised a way to make balloons airtight, by dissolving rubber in turpentine and painting the resulting varnish on silk. This actually ruined the look of their balloon—the red-and-white-striped envelope they'd envisioned came out looking red-and-stained-smoker's-teeth instead—but it did reopen the possibility of using hydrogen in balloons, since the tiny H_2 molecule couldn't diffuse through the varnish. A chemist the brothers knew, Jacques-Alexandre-César Charles, then solved the other problem with hydrogen by figuring out how to mass-produce it. He did so by filling a huge barrel with iron and dumping in some sulfuric acid (H_2SO_4). The acid decomposed when it struck the metal and released billows of hydrogen gas, which he fed into the balloon via fat leather tubes. It took him 1,100 pounds of iron and four messy days of work in August 1783, but he managed to fill the balloon with 34,000 liters of hydrogen. (To defray the cost of the work, he sold tickets to the filling, making it a public spectacle much like Lavoisier's experiments.) When launched, the balloon landed an incredible fifteen miles away in a field outside Paris. Peasants immediately attacked it with pitchforks and scythes, convinced that a monster had fallen from the sky. But once again, the destruction of a balloon couldn't tarnish the sense of triumph.

Peasants attack a balloon that landed outside Paris.

Stunned by the news, the Montgolfiers quickly shifted operations north to Versailles and began planning a second flight. Naturally, being French, they hired a local wallpaper designer to snazz up the look of their balloon. He crafted a stunning azure envelope studded with gold zodiac signs. This time the brothers also decided to send up test animals—a sheep, a rooster, and a duck; they dangled beneath the balloon in a cage. (You wonder why they bothered with a duck, which could fly on its own, but never mind.) They had the old shoe-burning routine down to a pseudoscience by that point, and while King Louis XVI looked on, they got their balloon inflated in just minutes. Then the fifteen men holding it to the earth let go. The crowd gasped as it lurched sideways; the animals squawked and bleated. But the balloon righted itself as it rose, and it eventually touched down, unharmed, two miles away.

The Montgolfiers and the Roberts now began scrambling for the real prize—the first manned flight. King Louis suggested sending up two criminals for this; he apparently took a dim view of anyone's chances of surviving. But both teams howled at the very idea; they didn't want convicts stealing their glory.

Based on scientific acumen alone, the Charles-Robert team surely had the advantage. The brothers were better engineers, and Charles was a superior chemist. But in the end enthusiasm beat book-learnin', and on November 21, 1783, the Montgolfiers helped two friends of theirs—a physicist named Jean-François Pilâtre de Rozier and a local marquis—into a gondola beneath their newest, orange balloon. The men holding the ropes let go, and the orange sphere began soaring across Paris like a slow meteor. Before that day surely billions of people throughout history had gazed at the sky, at the birds sweeping overhead, and thought, *Once, just once...* Pilâtre de Rozier and the marquis finally achieved this dream, landing five miles away and becoming the first human beings to fly.

Although the Charles-Robert team lost the race, Paris showed no less enthusiasm for their first manned flight two weeks later, in early December. By some estimates half the population of Paris—a half-million people—watched Jacques Charles and Anne-Jean Robert climb aboard. (Among the throng was ambassador Benjamin Franklin, who would sign the treaty ending the American War of Independence a week later. When a cynical spectator asked Franklin that day, "What use is a balloon?," Franklin murmured, "What use is a newborn baby?") Because hydrogen provides more lift than hot air, Charles and Anne-Jean remained aloft for two hours, landing twenty miles distant just after dusk. Like a kid at the fair, Charles immediately demanded another ride, and a few minutes later made the first solo ascent—rising 10,000 feet, high enough to watch the sun set a second time. Unfortunately the ups and downs of flying left him with a stabbing pain in his ear canals. After touching down, he never flew again—though whether that was because he knew he couldn't top the double sunset or because he didn't want to face the double earache again, we don't know.

Ballooning quickly became a public spectacle in Europe, albeit a dangerous one. Professional "aeronauts" began competing to see who could fly the highest, the farthest, the fastest. The most successful ones became bona fide celebrities, the Evel Knievels of their day. Other pilots started passenger balloon services, complete with picnic lunches: people dined on roast chicken, croissants, and iced lemonade while they gawked at the ground below. (Champagne proved a bust at high altitudes, though: due to the lower air pressure, the bubbles in the bottle expanded too quickly and whooshed out.*) On most rides, the only sour note for passengers was the cold temperatures aloft, which forced them to bundle up in furs and blankets. Every so often, though, balloons faced real peril. Weather patterns shift unpredictably at high altitudes, and crews sometimes got shelled with hail or exposed to lightning. Some people also suffered from oxygen sickness: blurry eyesight, Gumby legs, blackened fingers. (In extreme cases, people bled out of their eyeballs like Ebola victims.) And people began dying in balloon wrecks as early as 1785. In fact the first balloon fatality was the first man ever to fly, the physicist Pilâtre de Rozier, whose balloon caught fire over Normandy during an attempt to cross the English Channel. His fiancée, Susan, watched him drop half a mile and get dashed on the rocks below; she herself died from shock eight days later.

Besides the daredevils and picnickers, several scientists took to the sky as well. Some measured dew points and magnetic fields and other miscellanea. Others ran crude experiments that involved dropping booze bottles overboard and timing how long they took to smash onto the ground. Most contented themselves with simple, if surprising, observations. Flocks of butterflies might suddenly appear several miles up and hitch a ride on the balloon envelope. The acoustics of heaven also startled them. Because balloons drifted along with the wind, passengers heard no roar about them as they flew. As a result, sounds from the ground below were remarkably clear — cocks crowing, hammers pounding anvils, shotgun blasts. The sky above looked different at great heights, too. With less air to distort their light, stars looked less soft and twinkly, more like hard specks of ice. And during the day, the gentle blue of the sky gave way to a darker, more sinister Prussian blue.

Balloon flights also stimulated interest in the properties of gases. Indeed, it's no coincidence that some of the first balloonists were also some of the first people to explore the behavior of gases in a systematic way. These men had the most to lose, including their lives, if their theories failed.

One early question involved how exactly balloons got off the ground. The key principle here dates back to Archimedes, the famous streaker of ancient Syracuse. In the third century BC, the king of Syracuse gave a local goldsmith a hunk of gold to make a crown, but he suspected that the smithy had cheated him by keeping some of the gold and mixing in an equal weight of silver to replace it. Still, the king had no proof, so he turned to Archimedes for help. After puzzling over the assignment for weeks, Archimedes was lowering himself into a bath one day when he noticed the water level in the tub rising as he did so. An experiment flashed before his eyes, and moments later, Greek mothers were covering their children's eyes in the streets.

In most versions of this story, Archimedes solves the riddle by lowering the crown into a vessel half filled with water and noting how high the water rose. He then repeats the process with a nugget of pure gold that weighed the same amount. If the water rose to a different height in each case, then the goldsmith had cheated. But many historians dispute this scenario, since the difference in height between case one and case two would have been, in a typical Greek jug, around a half millimeter, far too fine to judge with the naked eye. (Try it.) Instead, Archimedes probably did this: He knew that water provides a buoyant upward force on anything submerged within it. (You feel this force when swimming.) And the bigger the volume of the object, the more upward force the water provides. So Archimedes found a scale, balanced both the crown and a hunk of pure gold on it, and lowered the whole shebang into water. The crown obviously weighed the same as the nugget (which is why the scale balanced initially). But if the crown contained silver—a less dense metal than gold—the crown's volume would be greater. With a greater volume, the crown would feel more buoyant force underwater. And with more buoyant force on one side, the scale would no longer balance. Sure enough, when Archimedes submerged the scale, the crown side drifted upward. Eureka—the goldsmith had cheated.

Experiments like this gave rise to what's now called Archimedes' principle. It says, first, that any object submerged in a fluid feels a buoyant upward force; and second, that the greater the object's volume, the more force it experiences. (More precisely, the upward force equals the weight of the fluid displaced.)

So what does all this have to do with balloons? Seemingly little—until scientists in the 1700s realized that air is also a fluid and also exerts a buoyant upward force. You don't feel this force in daily life and start floating away because it's fairly small and because your body is so compact. But it does exist: you actually weigh a smidgen less in air than you would in a vacuum. Large, less dense bodies like balloons feel this buoyant force acutely. Balanced against this force is of course the weight of the balloon and any cargo attached to it (like a basket), which wants to drag it down to earth. There's also the weight of gas itself to consider. That is, when you're calculating whether or not your balloon has enough lift to fly, you can't disregard the weight of the lifting gas inside the envelope. (It may be helping to generate lift, but it's not "free" payload; it still counts as weight.) That's why light gases like hydrogen are so useful in balloons. Because they weigh so little, gravity can't get as much of a purchase on them; the buoyant force therefore has a better shot at outmuscling gravity.

(Hydrogen, as the lightest element, provides the most lift per molecule. But really any gas lighter than air can lift a balloon—helium, steam, ammonia. In theory even a vacuum would work. This idea founders on the fact that no known material is both strong enough to keep a vacuum balloon from crumpling inward and yet light enough for air to lift it—copper balloons don't get too far.)

Although more popular early on, hydrogen balloons ultimately proved too expensive, too flammable, and too unwieldy for everyday use. (Hydrogen provided so much lift that even pouring a glass of water over the side, or peeing, could cause the balloon to lurch upward.) Most aeronauts therefore turned to hot-air balloons—and discovered in the process that hot-air balloons rise for slightly different reasons than hydrogen balloons do. Archimedes plays a role here, too, but we need some additional tools, the gas laws, to fully understand what's going on.

By the later 1700s scientists knew a bit about gas laws, which describe the interrelationship between gas temperature, gas pressure, and gas volume. For instance, Irish chemist Robert Boyle determined in 1662 that raising the temperature of a gas increases its pressure. For hot-air balloons, we need to look at volume versus temperature. One fundamental rule of gases is that heating them causes them to expand; contrariwise, cooling them causes them to contract. You can see this for yourself with a latex balloon: pop it into the fridge, and it shrinks; pull it back out, and it expands.

With hot-air balloons, something slightly different happens. The air inside them still expands when it heats up, of course. But unlike latex balloons, hot-air balloons are usually made of materials that don't stretch. As a result, the balloon itself won't expand much. Instead, the expanding gas pushes out of the hole in the bottom of the balloon, since it has nowhere else to go. This leaves less gas inside the envelope and therefore less weight. With less weight to lift, the buoyant upward force of Archimedes can now outmuscle gravity and hoist the balloon into the blue.

Ironically, even though the Montgolfiers built hot-air balloons and Jacques Charles built hydrogen balloons, it was Charles who first discovered the temperature-volume law for gases and thereby explained how hot-air balloons worked. He did much of the research that led to this discovery in his laboratory in the Louvre. For some reason, however, he hesitated to publish his temperature-volume law, so he generally receives little credit for it today. But he did discuss his experiments with a colleague, Joseph-Louis Gay-Lussac, who expanded on and refined Charles's experiments and eventually published the law himself in 1802.

Like Joseph Montgolfier, Gay-Lussac's life changed because of lingerie. In his case Gay-Lussac met his wife in a lingerie shop; she was the only clerk, remarkably enough, reading a chemistry text on the job. She in turn fancied him because, like many of his fellow chemists, Gay-Lussac had some cowboy in him. Once, while isolating sodium and potassium, the metals exploded in his face, nearly blinding him. He later shocked himself so badly with a homemade battery that he lost the use of his arms for a full day. Another flask blew up in his face in 1844; this time he definitely would have lost an eye, except he was wearing glasses, a result of his first mishap.

Naturally, Gay-Lussac loved risking his life in balloons and took to the skies whenever possible. For one celebrated flight in August 1804 he teamed up with a physicist named Jean-Baptiste Biot. Although technically copilot, Biot wasn't much use in the basket and spent most of the flight measuring Earth's magnetic field at different altitudes. (To his delight, he found that it remains strong even several miles high. Unlike temperature or pressure, it barely changes with elevation.) Gay-Lussac, meanwhile, hoped to gather data on the chemical composition of the upper atmosphere. To this end he planned to open several evacuated flasks at different altitudes and bottle the air there for later analysis.

As they continued to climb, however, things went awry. Biot wasn't exactly in miler's shape, and he began to feel woozy. Suddenly he slumped over as if dead. In the end he was fine, but you can imagine Gay-Lussac's panic. And between flying the balloon and resuscitating Biot, Gay-Lussac forgot about his flasks until they touched down. The missed opportunity irked him, so he made a solo ascent three weeks later. This balloon had more lift, and without Biot's ballast, he shot up to 23,000 feet, an altitude record that stood for half a century. He almost certainly suffered from oxygen debt at that height, but he shook off the mental torpor and managed to fill several flasks with rarified, high-altitude air. Analysis later revealed that even several miles up, the composition of the atmosphere remains the same as at sea level. In other words, while there's less air overall up high, the relative proportions of nitrogen, oxygen, and other gases is identical.

As atmospheric chemists began to push higher and higher, however, using unmanned balloons, they realized that the air does change in other, subtler ways as you ascend. They eventually discovered four distinct layers in Earth's atmosphere, each one surrounding the next like the layers of an onion. We live in the troposphere, which extends from the ground to around ten miles high (depending on the latitude and season). Most weather occurs here. Next comes the stratosphere, which contains ozone and reaches thirty-some miles high. The mesosphere, where meteors burn up, extends to around fifty miles. Finally you encounter the wispy thermosphere, which houses the aurora borealis and extends up to five hundred miles. As an altitude, five hundred miles seems vertiginously high—touching the edge of outer space. On the other hand, it's still just five hundred miles; you could drive your antigravity car there in about seven hours. (The scenery on this drive would get pretty bleak pretty fast, though: fully half the weight of Earth's atmosphere lies below four miles.) And at just seven miles up—no farther than a Sunday jog— oxygen levels have dropped to roughly one-quarter of their value at sea level, so low that life all but ceases. Scientists in ancient Greece argued that the air must improve as you climb toward the heavens— you reach quintessence. In truth, we live inside a precariously thin shell of air, far less thick proportionally than the skin on an apple.

In addition to exploring the atmosphere, Gay-Lussac put in enough lab work to discover yet another gas law. This one says that if the pressure of a gas increases, the temperature must increase as well. And if you're having a bit of déjà vu here—that's basically the same as the volume-temperature law—you're not the only one. Chemists in Gay-Lussac's day began to notice the same variables (temperature, pressure, volume) popping up over and over whenever they talked about gas behavior. Naturally, they began to wonder if they couldn't maybe combine a few individual gas laws into one overarching rule—the über law of gases. It took another generation, but in the 1830s, scientists finally discovered the so-called ideal gas law. Get ready. If you let V stand for volume, T for temperature, P for pressure, n for the amount of gas present, and R for a constant (to make the arithmetic balance), then you can describe the behavior of any known gas as follows: $PV = nRT$.

Now, that formula might not look impressive on the face of it. (You're hardly wishing you had sunglasses right now, I bet.) But believe me when I say that it's nitroglycerin: an incredible amount of power packed into a tiny space. When wielded properly, this law can tell you all sorts of things about gas behavior in an instant: how changing the temperature affects the volume, how changing the volume affects the pressure, or *any other combination* you can think of. Equally important, you can pluck hidden knowledge out of this equation. Let's say that the pressure gauges on a vat in your factory seem to be giving the wrong readings, to the point that you're worried the vat might explode. How can you tell? Well, measure the vat's volume, temperature, and so on, and voilà—the ideal gas law reveals the pressure. In other words, if you know the values of the other variables, the pressure falls out automatically. Or if your thermometers start to malfunction, you can infer temperature for free—or whatever else. It seems like cheating on some level: you're measuring one thing and getting something totally different, as if you could learn the color of something by measuring its height. But however dissimilar they seem from the outside, a gas's pressure, volume, and temperature are all related on a deeper, more fundamental level—and the ideal gas law explains how.

Indeed, the gas law reveals a whole new way of thinking about gases. Priestley and Lavoisier and Davy had focused on the differences between gases—how each new gas smelled different or burned differently or affected our physiology in different ways. But the ideal gas law applies to all of them equally. It banishes all talk of differences—it's democratic. A cloud of hydrogen is made of gnat-sized molecules, a cloud of radon of huge bumblebees. Yet both expand the exact same way if you heat them. Clouds of nitrogen gas are serene and nonreactive, clouds of chlorine gas searingly toxic. Yet if you increase the pressure on either, they contract by the exact same amount. Different solids and liquids share very little in common, beyond the obvious fact that they're usually hard or wet. But gases mimic one another to an incredible degree. Physically if not chemically, all gases behave the same way.

One of the highest callings of the human mind is scientific discovery: to look at the booming, buzzing confusion of the world around us and distill some sort of unchanging essence. $PV = nRT$ achieves that to a degree that few other scientific principles can. Every gas you can think of is represented in those five letters. Old-timers sometimes call the ideal gas law the perfect gas law, and I'd love to revive that name. Because the law really does hint at something absolute and ideal, something eternal and undying, something truly perfect at work in the world.

—◁◦▷—

Okay, time for a confession: that last section ended with a lie. A white lie, a well-intentioned lie, a cute and charming and maybe even a noble lie, but a lie nonetheless. Because when you're talking about real, actual gases—like the air you're breathing right now— those gases don't quite live up to the perfection of the perfect gas law. It's like how any circle you draw, even with a compass, won't quite be as round and symmetrical as an ideal circle. Chemists know this, of course, and when they calculate something with the ideal gas law, they expect that reality will deviate a tad. Still, some gases get closer to perfection than others. In fact, when two scientists discovered a few pretty-much-ideal gases in the 1890s—gases as close to perfect as anything in this world can get—the rest of the scientific community refused to believe them. Such flawless specimens, they thought, simply could not exist.

This story starts with the man who had maybe the highest tolerance for tedium in the history of science. John William Strutt was born rich and sickly in 1842, and should have led the useless life of a marginal English peer. He had to drop out of Eton due to whooping cough, and after he resumed schooling he hardly distinguished himself. But there was something ambitious in Strutt. He enrolled at Cambridge University, and rather than treating it like a finishing school, he horrified his family by taking math and physics classes. His success in the classes further stupefied them, and they were gobsmacked when he deigned to take *a job* there as a physics fellow. He appeared to come to his senses in 1871 when he married and resigned his post. In 1873, on the death of his father, Strutt became Lord Rayleigh ("Ray-lee") and took over the family estate. But that old itch needed scratching, and Rayleigh eventually turned the estate over to his younger brother in 1876. In other words, Rayleigh brought down upon himself the same humiliation that Joseph Montgolfier had endured—a younger brother taking over the family business—just so he could focus on science.

Rayleigh wore a walrus mustache and bushy sideburns, and year by year the balding dome of his head grew more pronounced. He published over four hundred papers in his career and did pioneering work in a dozen fields, including biology. (Among other things, he discovered why peacock feathers have that sexy metallic sheen.) But he took on a real snoozer of a topic in the early 1880s and began measuring the densities of gases like oxygen and hydrogen. To be sure, this work had been important once—in the 1700s. But a century on, it wasn't clear what Rayleigh expected to learn. He nevertheless spent ten years off and on doing this work—and did, in fact, learn nothing new. He simply extended the measurement of these gases' densities by a few decimal points. It was the un-eureka.

Unbelievably, masochistically, Rayleigh then began another

round of density measurements with an even more boring gas, nitrogen. To make the measurements he needed pure samples of nitrogen, and he got them by stripping out the other components of air one by one. To remove water vapor from the damp English air, he put wool blankets inside the reaction chambers in his lab; some days they soaked up two full pounds of water. He then ran the air through glowing-hot copper tubes to purge the oxygen, and through potash (a potassium-rich mineral) to remove the carbon dioxide. Sulfuric acid baths removed any last, trace impurities. He then collected the pure nitrogen in a bulb and measured its density.

Nothing if not obsessive, Rayleigh repeated the experiment by purifying nitrogen another way. Instead of removing the oxygen with glowing hot copper, this time he bubbled it through liquid ammonia. This was a bit messier chemically. Among other things, the O_2 reacted with the ammonia (NH_3) to produce extra nitrogen molecules as a by-product. But since Rayleigh wanted to study nitrogen anyway, adding more of it shouldn't have made a difference.

Guess what. Rayleigh found that a liter of nitrogen purified with ammonia was seven milligrams lighter than a liter of nitrogen purified with copper. That's 1/5,000th of an ounce. Having spent many futile hours in chemistry labs in college, I can tell you that if I'd ever gotten two trial runs to agree to within 1/10th of a pound, I would have wept with joy. You just don't get results that good—you'd investigate that person for faking data. But the difference tortured Rayleigh. Scientists in his day had scales a million times more sensitive than Lavoisier's had been, accurate to tiny fractions of a milligram. Seven milligrams therefore fell well within the experimental error, and Rayleigh began meticulously (neurotically) checking his results. He cooked up six more methods for purifying air, some of which made extra nitrogen, some of which didn't—and was "disgusted," as he said, to see that the difference persisted.

Rayleigh finally published a letter in *Nature* in 1892 admitting

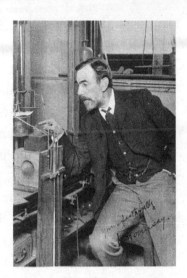

Lord Rayleigh (left), the physicist who discovered argon, and William Ramsay (right), the chemist who discovered additional noble gases. (Photos courtesy Wellcome Trust)

his bafflement and asking chemists worldwide for advice. Most of the suggestions he fielded came to naught. But he did listen to William Ramsay, a slender Scot with tired eyes who worked in London. Ramsay would soon get just as obsessed with the nitrogen discrepancy as Rayleigh was—a development that Rayleigh resented, feeling that Ramsay had butted in. They nevertheless agreed to keep each other abreast of their experiments and to publish them jointly.

One idea they ruled out immediately. Rayleigh had wondered whether he'd perhaps created an exotic form of nitrogen, like N_3 (similar to ozone, O_3). This could have explained the different densities, but Ramsay shot that idea down, since the structure of N_3 seemed unstable.

They finally got somewhere when they looked up an old, forgotten paper by Henry Cavendish. In 1785 Cavendish had trapped a pocket of air inside a tube filled with mercury. He then ran a current through the mercury, causing sparks to jump across the gap. These sparks caused the nitrogen and oxygen in the pocket to react, creating orange-red fumes. The fumes dissolved into the mercury, and as they did so the pocket shrank bit by bit. Cavendish expected that the pocket would eventually shrink to nothing, but no matter how long he ran the experiment—hours, days, weeks—around 1 percent of the gas refused to disappear. Eventually he gave up, but in reading the paper a century later, Rayleigh and Ramsay realized that he may have isolated a new gas.

Admittedly, this was a long shot. By 1892 scientists had been studying the composition of the atmosphere for more than a century, and it seemed unlikely—to say the least—that they'd missed a full 1 percent of the air. On the other hand, an extra gas would explain Rayleigh's funny densities at a stroke. Say that a gas heavier than nitrogen was indeed lurking in the scrubbed samples of air. The density of those samples would depend on the ratio of nitrogen to gas X, just as the density of a can of mixed nuts would depend on the relative proportions of peanuts to Brazil nuts. But whenever Rayleigh used a method that inadvertently added more nitrogen, more peanuts, to the mix, he would have thrown that ratio off and changed the density a smidge. Ramsey and Rayleigh agreed that looking for gas X was worth a shot.

In searching for gas X, each man adopted a different strategy. Rayleigh summoned the spirit of Cavendish and used electric sparks to purge the air of everything but the mystery gas. This work proved so monotonous that even Rayleigh grew weary of it. He set up a telephone line in his home lab and put one receiver near the spark generator, which hummed as it worked; he snaked the other receiver to his library, where he could snooze in a chair—and leap up whenever the humming over the line ceased. Ramsay meanwhile stripped the N_2 out of his sample by exposing it to magnesium, a superreactive metal that formed a crusty brown powder when it combined with nitrogen. Inducing this reaction, though, required heating the magnesium to the point that the glass tube around it almost started melting. Ramsay proved every bit the sucker for monotony that Rayleigh was, once spending ten days passing nitrogen back and forth across the hot metal. But both sets of experiments worked: each man acquired a thimbleful of gas X at 99 percent purity.

As a next step the duo investigated the properties of the gas. It proved tasteless and odorless and colorless, all intriguing signs. What really made the gas special, though, was its acoustics. Sound waves are basically pulses of energy spreading through gas molecules as they knock into one another. (It's like shoving someone in a crowd: he'll knock into other people, and they into other people, and so on, spreading the original energy of the shove outward.) This movement creates patterns of high and low pressure that our ears interpret as tones. And these wave-shoves move at different speeds depending on the weight and shape of the gas molecules involved.

Now, the physics here gets pretty gnarly (trust me). But in simple terms, Ramsay and Rayleigh measured the speed of sound in gas X under two conditions, then converted those two numbers into a ratio. Doing this would help determine the shape of the gas's molecules, since—and this is the key point—the ratio varies depending on the molecules' size and complexity. As an example, imagine sending a pulse of sound through a cloud of carbon dioxide (CO_2) or ammonia (NH_3). Within those molecules, the atoms of C and O or N and H can twist relative to each other or boing back and forth like springs. These extra gyrations end up dissipating some of the sound energy. And for other gnarly-physics reasons, this waste of energy lowers the crucial ratio to around 1.3. Simpler gases like H_2 and N_2 have fewer ways their atoms can move relative to each other and therefore waste less energy. This yields a higher ratio, 1.4. Ramsay and Rayleigh's mystery gas had a ratio of 1.67, meaning it was simpler than two atoms. In other words, it had one atom—unlike any other known gas.

That wasn't even the biggest surprise. Ramsay and Rayleigh also wanted to see how their gas reacted with other substances, so they exposed it to oxygen, hydrogen, and carbon dioxide. Nothing happened. They tried feistier compounds like sulfur, phosphorus, and potassium. All duds. They tried chlorine and acids and other horrors. It didn't blink. They eventually cleaned out their entire chemical supply closet, throwing every nasty substance they could at this gas. Not a flicker of interest.

By August 1894, the duo knew they had a new gas, and probably a new element. Still, they hesitated to discuss the possibility in public. To be sure, they reported the funny sound ratio and all the nonreactions—the bare results of their experiments—but they declined to go beyond that and stake a claim for a new element. Partly this was good, healthy circumspection—the need to nail everything down. Partly this was vanity. They'd recently discovered that the Smithsonian Institution was offering a prize of $10,000 ($275,000 today) for the best paper on an original discovery about air. Since the Smithsonian wanted only unpublished results, the duo had to stay mum so they could win the prize.

The media embargo didn't apply to their peers, though, who knew darn well what the results implied and didn't like it. Atmospheric chemists felt outright insulted: how could *all* of them have missed this mystery gas? (If gas *X* made up 1 percent of the atmosphere, as the duo claimed, then every human being alive was inhaling a third of a pound of it every day.) Worse, if R&R really had discovered a new element—which some scientists were now referring to as argon, after the Greek word for "lazy"*—it needed a slot on the periodic table. But where? Based on the density of argon, Rayleigh had calculated its atomic weight as 40. But that put it near chlorine and potassium on the table—two notoriously reactive elements. It made no sense. The father of the periodic table, Dmitri Mendeleev, finally weighed in and dismissed argon as ridiculous, suggesting that they'd somehow created N_3 instead.

Rayleigh and Ramsay won the Smithsonian prize in January 1895, beating 218 contestants. It proved a Pyrrhic victory, since scientific consensus had hardened against them during their silence. One critic denounced argon as a "chemical monster, brought unexpected and unwelcome, like the cuckoo, into the previously happy family of the elements." The periodic table had become so vital to chemistry that anything that threatened the table's legitimacy threatened chemistry itself. Rayleigh and Ramsay ended up apologizing for all the chaos — but without retracting a syllable.

Ramsay then went and made things worse with another discovery. A few years before, a naïve geologist in the United States had been working with some uranium ore when he noticed tiny bubbles of gas streaming out of it. The gas didn't react with anything he had at hand, so he called it nitrogen and moved on. After the debut of argon, a friend tipped off Ramsay to this story. Ramsay collected some of the rock fizz himself and confirmed that it didn't react with anything — not even things that normally combined with nitrogen. He then went one step further and showed that this gas weighed far less than argon — meaning Ramsay had scored another new element. He named it krypton, from the Greek *kryptos,* for "hidden, secret." But when he sent a sample to another scientist for confirmation, it was Ramsay's turn for a shock. As far back as the 1860s astronomers had been breaking down sunlight into its constituent colors. And among these colors they noticed odd bands of yellow and green and red sometimes, which they attributed to a mystery element. Ramsay's new gas, when heated, produced the exact same colored bands. In other words Ramsay had simply rediscovered this "sun element" on Earth. Because the earlier scientists had already named it, Ramsay bowed to precedent and called his gas helium.*

Helium compounded the problem posed by argon, because chemists now had to find room for two cuckoos in the harmonious nest of the periodic table. Although a few brave souls suggested adding a new, gas-only column to the table to accommodate them, Mendeleev and others scorned the very idea. The periodic table was arguably the most important breakthrough in chemistry history. Hundreds of scientists had spent millions of hours refining it. And now they were supposed to rearrange everything and shoehorn in a new column because of two bloody Brits? Hardly.

But Ramsay liked the idea of a new column—in part because it implied that more of these aloof gases existed. To test this hunch, he and an assistant began stripping down 1,600 liters of air in 1898, removing the oxygen, nitrogen, and other components until only nonreactive gas remained. They then cooled the leftover gas down hundreds of degrees into liquid form. They knew this liquid was mostly argon, but if other elements were lurking inside it, they could fish them out by slowly heating it. As we saw before, different substances within a liquid all boil off independently, at different temperatures. Sure enough, three new gases popped out during the slow crawl toward room temperature. For one gas, they recycled the name krypton. Another one they named xenon, after the Greek for "stranger." (Several of these gases have vaguely insulting names.) For the third gas, Ramsay announced its discovery over dinner one night with his family. His ten-year-old son Willie cut in and rather boldly proposed a name for it, *novum,* Latin for "new." Dad chewed on this overnight, but he wanted consistency with the other Greek etymologies, so they compromised on neon.

Now, you might think that five cuckoos were even worse than two, but no. Given five nonreactive gases—argon, helium, neon, krypton, and xenon—chemists felt far more comfortable opening up a new, gas-only column on the periodic table. (Years later scientists would discover a sixth gas for this column, radon. Ramsay then confirmed radon's existence by examining its light spectrum—which meant that Ramsay had a hand in discovering every known noble gas,* an unprecedented feat.) To be blunt, many chemists treated this new column like a ghetto at first, a place to dump these problematic gases and forget about them. In time, though, most learned to appreciate them; even cranky old Mendeleev smiled.* Nowadays these gases are revered as the "noble gases," since they don't deign to interact with other elements. They're perfectly content to remain alone, and this lack of interest in other atoms allows them to obey the perfect, ideal gas law to an extraordinary degree.

Ramsay ended up winning one of Alfred Nobel's new chemistry prizes in 1904, and he became a scientific celebrity. In interviews he modestly attributed his success to his fat thumbs, which he'd used to stopper glass tubes when transferring noble gases around his lab. He also credited the fine-motor skills he'd developed from years of rolling his own cigarettes. (He disdained store-bought ones as "unworthy of an experimentalist.") Eventually all that smoking caught up with him, and he developed nasal cancer in the mid-1910s. During his decline, he grew obsessed with World War I and the seeming end of civilization, and he began lashing out at German scientists with vitriolic letters in *The Times* of London. (His friends, embarrassed by his conduct, explained this obsession with Germany as temporary insanity brought about by the pain of the cancer.) Ramsay died a bitter man in 1916, crushed to learn that human beings seldom behave as perfectly as his gases.

Meanwhile, Rayleigh won the Nobel Prize in Physics in 1904 for discovering argon. The award was meant to complement Ramsay's chemistry award, but given all the exotic new gases Ramsay had discovered, Rayleigh's prize seemed like an afterthought. Indeed, you might be wondering what the heck Rayleigh had been doing for the previous decade — just standing by while Ramsay rewrote the periodic table? Not quite. Rayleigh had followed his own inclinations, and in a remarkable feat of physics he'd finally proved, after millennia of speculation, why the sky is blue.

Before 1900 philosophers and protoscientists had come up with all sorts of explanations for why the sky is blue. Some claimed that the color represented a compromise hue, a blend of the indigo of the night with the yellow of the sun. Others attributed it to floating ice crystals; still others to exotica like ozone fluorescence or microscopic bubbles. For his part Rayleigh traced the blue to the scattering of sunlight by unknown particles in the air. Again, the physics here gets hairy, but the important point is that light with shorter wavelengths gets scattered (i.e., redirected) far more than light with longer wavelengths. In particular, blue light, which has a shorter wavelength than almost any other color in the rainbow, gets scattered far more easily than red or orange light.

This scattering leads to a blue sky as follows. Imagine you're lying on a blanket outdoors, watching clouds drift by. Meanwhile some white light from the sun arrives at Earth. In truth this "white light" is made up of several different colors of light, including blue, layered on top of one another. According to Rayleigh's theory, there's a higher chance of that blue light getting scattered and diverted than of any other color being redirected. Now, this blue light could fly off in any direction after this happens. It might get pushed seventeen miles north of you. It might get redirected upward, into space. But some blue light will be redirected down from the sky into your pupil. To be sure, some red (or yellow or green) light might also get redirected your way. But far less of it than blue light; at any one point in the sky, blue dominates. Multiply this one point by trillions of trillions of others, and you get a beautiful azure sky.

(Those of you who know your color theory might be ready to throw the bullshit flag here. After all, purple light has an even shorter wavelength than blue light. So by the reasoning above, the sky should be periwinkle! That's true so far as it goes, but it overlooks other factors. The sun happens to emit more blue than purple light, so there's more blue light to scatter. Furthermore, the cones in our retinas can't detect purple light all that well. A full explanation of the blue sky, then, takes into account not only Rayleigh's scattering but the sun's spectrum and the circuitry of our eyes.)

Rayleigh laid out much of this explanation in 1871, the year he traded Cambridge for marriage. But in doing so he glided over one crucial point: the identity of the particles that actually scatter sunlight. Was it dust, ice, fleets of airborne microbes? Rayleigh suggested salt crystals, but no one really knew.

The germ of the answer first appeared in a letter Rayleigh received from physicist James Clerk Maxwell in 1873. Maxwell was vacationing in northeast India, and he'd just spent an afternoon on his hotel terrace gazing at the Himalayas. He could even make out Mount Everest a hundred miles distant. The clarity of the air stunned him, and he wondered why the gas molecules between him and Everest hadn't absorbed all the light. Had Maxwell put his mind to it, he surely could have solved the riddle; he was a pantheon physicist who'd already rewritten both thermodynamics and light theory. But Maxwell didn't have any physics books handy to get started, and he admitted in his letter that he was feeling a little lazy anyway, so he deputized Rayleigh to investigate.

Rayleigh took his sweet time with the assignment—a quarter century passed, and in the meantime Maxwell died. But in 1899 Rayleigh finally made the necessary calculations. As a corollary of that work, he found that air molecules are the perfect size to scatter visible light. So it wasn't impurities like dust or salt or bubbles at all. "Even in the absence of foreign particles," Rayleigh declared, "we should still have a blue sky." Nitrogen and oxygen and argon alone* paint the dome of heaven blue. And Rayleigh never would have known this had his old friend Maxwell not spent a slothful afternoon staring at Mount Everest, the point on Earth that comes closest to touching that sky.

————◄○►————

Chemists in the 1780s fulfilled maybe the oldest dream of humankind, to snap the tethers of gravity and take flight. A century later a physicist solved one of humankind's most enduring mysteries, why the sky is blue. So you can forgive scientists for having a pretty lofty view of themselves circa 1900, and for assuming that they had a full reckoning of how air worked. Thanks to chemists from Priestley to Ramsay, they now knew all its major components. Thanks to the ideal gas law, they now knew how air responded to almost any change in temperature and pressure you could throw at it. Thanks to Charles and Gay-Lussac and other balloonists, they now knew what the air was like even miles above our heads. There were a few loose ends, sure, in fields like atomic physics and meteorology. But all scientists needed to do was extrapolate from known gas laws to cover those cases. They must have felt achingly close to figuring out their world.

Guess what. Scientists not only ran into difficulties tying those loose ends together, they eventually had to construct whole new laws of nature, out of sheer desperation, to make sense of what was going on. Atomic physics of course led to the absurdities of quantum mechanics and the horrors of nuclear warfare. And tough as it is to believe, meteorology, one of the sleepiest branches of science around, stirred to life chaos theory, one of the most profound and troubling currents in twentieth-century thought.

Interlude: Night Lights

Acetylene (C$_2$H$_2$)—currently between 0.0001 and 0.001 parts per million in the air (or higher in urban areas); you inhale between a billion and ten billion molecules every time you breathe

Hydrogen (H$_2$)—currently 0.55 parts per million of air; you inhale seven quadrillion molecules every time you breathe

We can't leave the noble gases behind without talking a little bit about their most famous application, in so-called neon lights. To be fair, though, neon lights are actually part of a much bigger story about light and gases in general. Because while gases like steam certainly powered the Industrial Revolution, gases like methane and acetylene did something just as important: they lit the revolution.

One historian has even called such light-giving gases, along with steam, "the two main driving forces in history."

For some context here, recall that Joseph Priestley's Lunar Society met on the Monday nearest the full moon because its members needed moonlight to find their way home. But Priestley's generation was among the last to have to worry about such problems. Several of the gases that scientists discovered in the late 1700s burned with uncanny brightness, and within a half century of Priestley's death in 1804, gas lighting had become standard throughout Europe. Edison's lightbulb gets all the historical headlines, but it was coal gas that first eradicated darkness in the modern world.

Human beings had artificial lighting before 1800, of course — wood fires, candles, oil lamps. But however romantic bonfires and candlelit dinners seem nowadays, those are actually terrible sources of light. Candles especially throw off a sickly, feeble glow that, as one historian joked, did little but "make darkness visible." (A French saying from the time captured the sentiment in a different way: "By candlelight, a goat is ladylike.") Not everyone could afford candles on a daily basis anyway — imagine if all your lightbulbs needed replacing every few nights. Larger households and businesses might go through 2,500 candles per year. To top it off, candles released noxious smoke indoors, and it was all too easy to knock one over and set your house or factory ablaze.

In retrospect coal gas seems the obvious solution to these problems. Coal gas is a heterogeneous mix of methane, hydrogen, and other gases that emerge when coal is slowly heated. Both methane and hydrogen burn brilliantly alone, and when burned together, they produce light dozens of times bolder and brighter than candlelight. But as with laughing gas, people considered coal gas little more than a novelty at first. Hucksters would pack crowds into dark rooms for a halfpenny each and dazzle them with gas pyrotechnics. And it wasn't just the brilliance that impressed. Because they didn't depend on wicks, coal-gas flames could defy gravity and leap out sideways or upside down. Some showmen even combined different flames to make flowers and animal shapes, somewhat like balloon animals.

Gradually, people realized that coal gas would make fine interior lighting. Gas jets burned steadily and cleanly, without a candle's flickering and smoking, and you could secure gas fixtures to the wall, decreasing the odds of things catching fire. An eccentric engineer named William Murdoch—the same man who invented a steam locomotive in James Watt's factory, before Watt told him to knock it off—installed the world's first gas-lighting system in his home in Birmingham in 1792. Several local businessmen were impressed enough to install gas lighting in their factories shortly thereafter.

After these early adopters, city governments began using coal gas to light their streets and bridges. Cities usually stored the gas inside giant tanks (called gasometers) and piped it through underground mains, much like water today. London alone had forty thousand gas street lamps by 1823, and other cities in Europe followed suit. (Paris didn't want bloody London usurping its reputation as the city of light, after all.) For the first time in history, human settlements would have been visible from space by night.

Public buildings came online next, including railway stations, churches, and especially theaters, which benefitted more than probably any other institution. With more light available, theater directors could position actors farther back onstage, allowing for more depth of movement. A related technology called limelight—which involved streaming oxygen and hydrogen over burning quicklime—provided even brighter light and led to the first spotlights. Because the audience could see them clearly now, actors could also get by with less makeup and could gesture in more realistic, less histrionic ways.

Even villages in rural England had rudimentary gas mains by the mid-1800s, and the spread of cheap, consistent lighting changed society in several ways. Crime dropped, since thugs and lowlifes could no longer hide under the cloak of darkness. Nightlife exploded as taverns and restaurants began staying open later. Factories instituted regular working hours since they no longer had to shut down after sunset in the winter, and some manufacturers operated all night to churn out goods.

Another gas provided the first strong portable lights. In 1836 Humphry Davy's cousin Edmund discovered acetylene, a hard little plug of hydrogen and triple-bonded carbon (C_2H_2). It burned with startling ferocity and quickly found use in street lamps, buoys, and lighthouses. Entrepreneurs also developed handy acetylene lanterns for portable use, especially in mine shafts and caves. In later years bicycles and cars, including the Model T, employed acetylene headlamps, despite one peculiar side effect. Most lamps and lanterns created acetylene by dripping water onto a brittle gray mineral called calcium carbide (CaC_2). The acetylene that bubbled off had no odor, but certain by-products of the process reeked like garlic.

Despite its advantages over candlelight, gas lighting wasn't a perfect technology. Coal gas sometimes released impurities like ammonia and sulfur that made people ill. The intense flames tended to gobble up the oxygen in a room, leaving people with headaches after a night at the theater. Gas fixtures sometimes leaked and asphyxiated people as well. Philosopher Friedrich Schiller notoriously praised the spread of gas mains as a quick, painless way to commit suicide—an "endorsement" that didn't help the less-than-wholesome reputation of gas lighting.

By the early 1900s most cities had started swapping out gas fixtures for lightbulbs, which didn't smell, didn't steal oxygen, and provided brighter light. Bulbs seemed more modern, too, the next logical step in a progression: coal gas had provided pure fire without wood or smoke; lightbulbs gave pure light without even the fire.

Still, lightbulb makers couldn't disregard gases entirely in their designs. The filament inside most bulbs consists of a thin strip of metal (often tungsten). Running electricity through the metal makes it glow but also heats it up, and in the presence of oxygen, hot metals burn. To eliminate this problem, manufacturers began pumping all the air out of lightbulbs, leaving a vacuum inside. Solving the one problem only introduced another, however, since hot metal filaments slowly evaporated in ultralow pressure, blackening the interior of the bulb. Most bulbs nowadays are therefore evacuated first and then refilled with nitrogen or another inert gas.

If lightbulbs eliminated flames, some modern lighting systems go one better and eliminate the filament as well. Take vapor lighting, the basis of yellow sodium streetlights. Vapor lighting differs from nineteenth-century gas lighting because gas lighting involved a chemical reaction: the methane and hydrogen molecules broke their internal bonds, released heat and light, and formed new substances. Vapor lighting does not involve breaking bonds or making new substances. Rather, you run electricity through a gas of sodium atoms and excite them. More specifically, you excite the electrons in the sodium atoms, which start leaping to higher energy levels and then crashing back down moments later. This jumping up and crashing down happens to release photons of light, which radiate outward and ping our eyeballs, helping us swerve around that pothole.

Neon lighting produces light via this same process of exciting electrons. To make neon lights, you can actually use any one of the six noble gases, depending on the color you want. Simply fill a tube with krypton or xenon or whatever, run a current through it, and shield your eyes. A French low-temperature chemist named Georges Claude sold the first neon advertising sign to a barber in Paris in 1912, followed by a rooftop billboard for vermouth. This led to a job lighting the entryway for the Paris Opéra, and further commissions from there. Neon lighting didn't exactly take off after that—Claude nearly went bankrupt in the 1920s—but he died in 1960 a rich, rich man. Oddly, many early computers and calculators used neon lights for their displays, since neon lights required less energy than traditional bulbs and didn't overheat as easily.

In the days before ubiquitous lighting, people sometimes called artificial light "borrowed light." It seems a quaint phrase now, as if you had to steal snatches of sunlight and smuggle them into the dark. Nowadays, of course, we worry more about keeping light *away* at night. We frown about light pollution ruining our view of the stars and fret over the street lamps outside our window disrupting a good night's rest. It's a turn of events that would have made our ancestors marvel: darkness itself has become a precious commodity in the modern world.

III. Frontiers

———◦———

THE NEW HEAVENS

A person living in the year 1600 wouldn't have felt all that out of place in the early 1700s. Even the early 1800s probably wouldn't have seemed too alien. But jump forward another hundred years, to the 1900s, and she would have been left agog. Steel skyscrapers suddenly towered above. Steamships had revolutionized trade and travel. Even the distinction between day and night, the most basic organizing principle in human life, had eroded. Gases played a major role in each advance, proving that air has shaped far more than our basic biology—it shaped human civilization as well.

In the past several decades the relationship between human beings and air has changed yet again. The air our friend from 1600 breathed is not the same air we breathe today; industrial development has changed its chemical composition. Our mental conception of air has changed even more dramatically: only recently have scientists started to appreciate just how complex our atmosphere is, rivaling the human brain in both its intricacy and its fragility. Until now, this book has focused on how the atmosphere has shaped human beings. Now we have to flip things around, and face up to how we human beings have shaped the atmosphere.

CHAPTER SEVEN

The Fallout of Fallout

Iodine-131 (I) and strontium-90 (Sr)—currently zero parts per million in the air (if you're lucky!)

It was already pretty weird to see a pig swimming in the South Pacific. And for the sailors who found her in the lagoon, it was all the more surreal considering they'd just obliterated the lagoon with a nuclear bomb.

Unlike the ultra-secret Manhattan Project, the U.S. military had been bragging about Operation Crossroads for months. Held in July 1946, at a cost of $100 million ($1.2 billion today), Crossroads was the biggest science experiment in history up until then—although "experiment" implies a level of refinement somewhat lacking here. The navy basically planned to drop an atomic bomb onto a fleet of ninety ships and just see what the hell happened. Crossroads nevertheless required 42,000 sailors to coordinate, plus 25,000 radiation detectors and 1.5 million feet of film—half the world's supply then—to gather data. The setting was the Bikini Atoll, a ring of coral islands 2,600 miles southwest of Hawaii. Paradise.

The navy had captured several of the target ships from Germany and Japan the previous year, including the hated *Nagato,* the command ship for the raid on Pearl Harbor. Controversially, the dummy fleet also included American ships that had fought in important campaigns, like the USS *New York* and USS *Pennsylvania.* (Officers did remove the bells and tea sets from each ship, which had sentimental value. They left the pinup girls.) After a public outcry, Congress stepped in and limited the number of American ships to thirty-three, but the fleet of battleships, dreadnoughts, and submarines in the Bikini lagoon still would have constituted the fifth-largest navy in the world.

Goats and rats exposed to fallout and radiation during Operation Crossroads. (Photo courtesy Getty Images)

More controversy erupted in the spring of 1946. First, the U.S. military evicted all 167 native Bikinians from the island and relocated them. The island's military governor had the chutzpah to call them lucky, comparing them to the Israelites being freed from bondage in Egypt and led into the promised land. (One guess who Moses was…) Second, in keeping with the biblical motif, the navy brought an ark's worth of animals to Bikini and distributed them among the target ships, to test the biological effects of atomic bombs. After this was announced, several thousand angry letters poured into U.S. government offices. Ninety people even volunteered to take the animals' places, including the writer E. B. White and a prisoner in San Quentin who said he wanted to do society some good for a change. (Some less-than-charitable folks suggested using prisoners of war in lieu of animals. Instead of Eichmann in Jerusalem, we'd have Eichmann in Bikini, which doesn't quite resonate.) The navy agreed not to use dogs, but did import 5,000 rats, 204 goats, and 200 pigs, among other creatures. To make the experiment more realistic, scientists dressed the larger animals in military uniforms the day before the test and cut their hair to human length. Pigs were chosen because they have organs that resemble human beings', while several of the goats had undergone conditioning* to make them prone to psychotic breakdowns; this would supposedly help determine the psychological effects of nuclear war.

The grand experiment began at 9:00 a.m. on July 1, 1946. A few minutes beforehand, thousands of sailors began lining the decks of the support ships outside Bikini lagoon. Officers ordered them not to look at the blast, but of course everyone did; the savvier ones watched with one eye closed, just in case. A joke making the rounds that morning had it that if the bomb obliterated Bikini, they could always change the name to Nothing Atoll (i.e., "at all"). Still, most men were nervous: in the eleven months since Nagasaki, no atomic weapons had been detonated, and the Bomb had acquired an almost supernatural aura in the public imagination.

Sailors' stomachs knotted as a B-29 bomber, *Dave's Dream,* appeared overhead. In its belly lay the nuke, which the flyboys had christened "Gilda," after bombshell Rita Hayworth's role in a recent flick. (Officially, the bomb test was known as Able, for A.) The bomb target, the USS *Nevada,* sat 3.5 miles from the Bikini shore, wedged in among a half-dozen other ships. The navy had painted *Nevada* hazard orange to make it easy to spot. The bombers missed anyway. Gilda detonated 520 feet over the water, as she was supposed to, but fell 650 yards northwest of *Nevada,* near an aircraft carrier.

The equivalent of fifty million pounds of TNT exploded in Gilda's bosom. The cyclopean sailors saw a flash of light and felt a warm flush on their cheeks; it took the roar two full minutes to reach them. The animals on the ships had less warning. The bomb vaporized everything nearby and sent a shock wave rocketing outward at 10,000 miles per hour. Many of the animals died from the concussive force, and five vessels within a thousand yards of the "zeropoint" started sinking. Every last animal aboard those ships was trapped and drowned in the wreckage. Except one.

Pig 311 — named for the numbered tag on her ear — had been dressed in a uniform and locked in the officer's head (toilet) aboard the Japanese cruiser *Sagawa,* 420 yards from the zeropoint. After the blast, it looked as if a giant had crushed *Sagawa* under his boot heel; the blast tore a hole in *Sagawa*'s side as well, and she began sinking. Somehow, though, amid the destruction, the toilet door popped open. And Pig 311 somehow got stripped naked and avoided being skewered as she scrambled through the wreckage and plunged into the lagoon. A patrol boat sent in to gawk at the damage the next morning (fallout, schmallout) found her frothing up the water, piggy-paddling for shore. She was six months old, mostly white with black patches, and weighed fifty pounds.

Despite the miraculous rescue, veterinarians took 311 for a goner.

She started losing weight and her hair fell out. More ominously, her blood-cell counts dropped, since radiation kills the bone marrow responsible for making new blood cells. Based on the symptoms in other exposed animals, her stomach and brain likely started to swell, and her liver likely started to atrophy—radiation sickness at its most acute. But somehow over the next few weeks she stabilized. Her hair grew back and her cell counts leveled off, then began to climb. Pretty soon she began gaining weight, and other vital signs sprang back, too. Before long she looked normal again.

The military, naturally, delighted in her recovery. Officials were eager to downplay the threat of nuclear weapons, and as soon as Pig 311 looked fat and happy again, they began promoting her as a folk hero—the little piggy who defied the big bad bomb. And the public bought the story hook, line, and stinker. *Life* magazine ran a photo spread, and a syndicated columnist declared her "a symbol of mind over matter and of pork over both." She soon landed a prized pen in the National Zoo in Washington, DC, where visitors from across the country lined up to see her. Some pig.

The Pig 311 propaganda (pro-pig-anda?) had one overriding purpose—to reassure the public about the safety of nuclear weapons. To a remarkable degree, it worked. True, when we look back on the nuclear age nowadays, we can't help but think of radioactive milk and children diving under desks. But our national nuclear freak-out didn't start right away. In the 1940s and early 1950s, people were just as likely to pooh-pooh nuclear weapons, even chuckle over them. Rather than Hiroshima and Nagasaki, they might think of Operation Crossroads and Pig 311. Because really, if a pig could survive an A-bomb, how bad could it be?

This complacency dovetailed nicely with the U.S. government's goal of testing as many nuclear weapons as possible, as quickly as it could. The military managed to squeeze in two hundred tests over the next two decades, and we're still living with the consequences today: even sixty years later, we're still inhaling some of the radio-active atoms these bombs gave birth to.

———◄○►———

The Manhattan Project wasn't a scientific breakthrough* as much as an engineering triumph. All the essential physics had been worked out before the war even started, and the truly heroic efforts involved not blackboards and eurekas but elbow grease and back-breaking labor. Consider the refinement of uranium. Among other steps, workers had to convert over 20,000 pounds of raw uranium ore into a gas (uranium hexafluoride) and then whittle it down, almost atom by atom, to 112 pounds of fissionable uranium-235. This required building a $500 million plant ($6.6 billion today) in Oak Ridge, Tennessee, that sprawled across forty-four acres and used three times as much electricity as all of Detroit. All that fancy theorizing about bombs would have gone for naught if not for this unprecedented investment.

Plutonium was no picnic, either. Making plutonium (it doesn't exist in nature) proved every bit as challenging and costly as refin-ing uranium. Detonating the stuff was an even bigger hassle. Although plutonium is quite radioactive—inhaling a tenth of a gram of it will kill most adults—the small amount of plutonium that scientists at Los Alamos were working with wouldn't undergo a chain reaction and explode unless they increased its density dra-matically. Plutonium metal is already pretty dense, though, so the only plausible way to do this was by crunching it together with a ring of explosives. Unfortunately, while it's easy to blow something apart with explosives, it's well-nigh impossible to collapse it into a smaller shape in a coherent way. Los Alamos scientists spent many hours screaming at one another over the details.

By spring 1945, they'd finally sketched out a plausible setup for the explosives. But the idea needed confirming, so they scheduled the famous Trinity test for July 16, 1945. Responsibility for arming the device—nicknamed the Gadget—fell to Louis Slotin, a young Canadian physicist who had a reputation for being foolhardy (perfect for bomb work). After he'd climbed the hundred-foot Trinity tower and assembled the bomb, Slotin and his bosses accepted a $2 billion receipt for it and drove off to watch from the base camp ten miles distant.

At 5:30 a.m. the ring of explosives went off and crushed the Gadget's grapefruit-sized plutonium core into a ball the size of a peach pit. A tiny dollop in the middle—beryllium mixed with polonium— then kicked out a few subatomic particles called neutrons, which really got things hopping. These neutrons stuck to nearby plutonium atoms, rendering them unstable and causing them to fission, or split. This splitting released loads of energy: the kick from a single plutonium atom can make a grain of sand jump visibly, even though a plutonium atom is a hundred thousand million billion times smaller. Crucially, each split also released more neutrons. These neutrons then glommed onto other plutonium atoms, rendered them unstable, and caused more fissioning.

Within a few millionths of a second, eighty generations of plutonium atoms had fissioned, releasing an amount of energy equal to fifty million pounds of TNT. What happened next gets complicated, but all that energy vaporized everything within a chip shot of the bomb—the metal tower, the sand below, every lizard and scorpion. More than vaporized, actually. The temperature near the core spiked so high, to tens of millions of degrees, that electrons within the vapor were torn loose from their atoms and began to roam around on their own, like fireflies. This produced a new state of matter called a plasma, a sort of übergas most commonly found inside the nuclear furnace of stars.

Given the incredible energies here, even sober scientists like

Robert Oppenheimer (director of the Manhattan Project) had seriously considered the possibility that Trinity would ignite the atmosphere and fry everything on Earth's surface. That didn't happen, obviously, but each of the several hundred men who watched that morning—some of whom slathered their faces in sunscreen, and shaded their eyes behind sunglasses—knew they'd unleashed a new type of hell on the world. After Trinity quieted down Oppenheimer famously recalled a line from the Bhagavad Gita: "Now I am become Death, the destroyer of worlds." Less famously, Oppenheimer also recalled something that Alfred Nobel once said, about how dynamite would render war so terrible that humankind would surely give it up. How quaint that wish seemed now, in the shadow of a mushroom cloud.

After the attacks on Hiroshima and Nagasaki in early August, most Manhattan Project scientists felt a sense of triumph. Over the next few months, however, the stories that emerged from Japan left them with a growing sense of revulsion. They'd known their marvelously engineered bombs would kill tens of thousands of people, obviously. But the military had already killed comparable numbers of civilians during the firebombings of Dresden and Tokyo. (Some historians estimate that more human beings died during the six hours of the Tokyo firebombing—at least 100,000—than in any attack in history, then or since.)

What appalled most scientists about Hiroshima and Nagasaki, then, wasn't the immediate body count but the lingering radioactivity. Most physicists before this had a rather cavalier attitude about radioactivity; stories abound about their macho disdain* for the dangers involved. Japan changed that. Fallout from the bomb continued to poison people for months afterward—killing their cells, ulcerating their skin, turning even the salt in their blood and the fillings in their teeth into tiny radioactive bombs.

General Douglas MacArthur, the military governor of Japan,

eventually declared a media blackout on all stories about the after-effects of the bomb. But MacArthur couldn't suppress the rumors, especially within the scientific community. The tellers of these tales didn't need to exaggerate, either—the reality was that bad. (One American doctor remembered walking into a relief station weeks afterward and wondering why all the Japanese patients were wearing polka-dot shirts. Then the polka dots began wriggling. The victims lacked the strength to brush away the insects snacking on their skin.) A series of hushed-up accidents in Los Alamos in 1945 and 1946 brought out the dangers of radioactivity even more acutely, since scientists could see the destruction firsthand.

The accidents started with Harry Daghlian, a roly-poly physicist who'd enrolled in MIT at age seventeen and had arrived in Los Alamos six years later, in 1944. He worked on criticality experiments, which involved starting chain reactions in spheres of plutonium in the lab. These experiments couldn't possibly have detonated the plutonium, since the spheres weren't dense enough. But the work wasn't exactly OSHA-approved, either. Scientists at Los Alamos referred to this research as "tickling the tail of a sleeping dragon," and it took place in a remote canyon four miles from the main Los Alamos grounds, in a facility called the Omega Site.

Daghlian worked with a 3.5-inch plutonium sphere identical to the bomb cores at Trinity and Nagasaki. Indeed, his sphere—nicknamed "Rufus"—would have found its way into a bomb around the twentieth of August if Japan hadn't surrendered and ended the war. Now, with the war over, you might think that Daghlian suddenly had nothing to do: the whole point of the Manhattan Project was to defeat Germany and Japan, after all, and that mission had been accomplished. But in truth, the workload for some people at Los Alamos didn't ease up much. The U.S. government had spent hundreds of millions of dollars on nuclear weapons and didn't want to waste that investment. More important, the scramble for postwar power had already begun, and several key figures within the government considered nuclear bombs—especially plutonium bombs—essential for long-term national security.

So on August 21, less than a week after the war ended, Daghlian began another series of dragon-tickling tests at the Omega building. To use the terminology of the lab, these tests involved creating a "nest" of blocks around a plutonium "egg." The blocks were made of a special material called tungsten-carbide that happens to reflect neutrons quite well. (Neutrons pass through many materials.) In stacking the blocks around the sphere, Daghlian therefore ensured that any stray neutrons that escaped would reflect back toward the plutonium, allowing them to participate in the fissionings. This effectively lowered the amount of plutonium needed to sustain a chain reaction. Daghlian completed one stacking test in the morning and a second in the afternoon, then attended a colloquium that evening. But after it ended, about 9 p.m., he returned to Omega to complete one last run. Working alone like this was prohibited, and Daghlian had already put in a long day, but Los Alamos was pretty informal then, and Daghlian was an affable guy, so the security guard let him in.

He first lowered the 14-pound sphere into a cradle. In addition to being radioactive, plutonium is toxic—almost as nasty as arsenic—so the sphere was candy-coated in nickel, making it "safe" to handle sans gloves. Scientists remember it feeling eerily warm from the radioactivity smoldering beneath the skin. One compared it to holding a live rabbit.

Egg in place, Daghlian began to assemble the nest of 52 tungsten-carbide reflectors. With each layer, more and more neutrons got volleyed back into the plutonium core, bringing it closer to criticality. With four full layers assembled and most of a fifth in place, he picked up one final brick and hovered it over the sphere. At this, a loudspeaker on the bench beside him started crackling. It was wired to a radiation detector, and the crackling meant that adding one last brick would start a chain reaction. Daghlian began waving the brick back and forth, teasing, toying, tickling. Satisfied with the chatter, he pulled back—at which point the brick slipped. It landed right where it shouldn't have, blocking the only escape hatch the neutrons had. In the jargon of the lab, the sphere went "super-prompt critical." More bluntly, Rufus the dragon opened its eyes.

The staccato of the speaker swelled into a rumble. An aurora of blue light began to dance around the sphere as the nitrogen in the nearby air ionized. Daghlian tried to knock the brick castle apart with his hand, failed, tried again, and finally toppled it. Everything went quiet. The aurora faded, the speaker hushed. Certainly no alarms sounded. And aside from a tingling in his hand, Daghlian felt fine: unlike a conventional poison, radioactivity causes no pain at first. But Daghlian had already killed himself, and he knew it.

He went to the hospital anyway. Doctors removed his keys and a knife from his pocket and told a technician to analyze them back at Los Alamos. She did, and found that they jammed the Geiger counters, they were that "hot." Physicists later estimated that his body had been blasted with the rough equivalent of 50,000 chest X-rays. His right hand got closer to 400,000, and over the next few days that hand turned into a gigantic blister. His right arm swelled, too, and the skin turned red and peeled off up to his shoulder, as did skin on his face. (He was essentially suffering from a "three-dimensional sunburn," the radiation had penetrated so deep inside his body.) Meanwhile, tsunamis of nausea washed over him, punctured by cramps and hiccups. His hair fell out by the handful, and whole pounds melted off his frame. Mercifully he slipped into a coma after a few weeks, and twenty-five days after the accident he died. Military officials lied to the press and said that Daghlian had succumbed to "chemical burns." They then wrote a $10,000 check to Daghlian's sister and mother to hush them up.

At Daghlian's bedside throughout the ordeal was Louis Slotin, the scientist who'd assembled the bomb for the Trinity test. Watching Daghlian die shook Slotin and reinforced a recent decision he'd made to abandon weapons work. Unfortunately, Slotin still had a year on his contract with the military. Even more unfortunately, he didn't learn anything from Daghlian's accident about being more careful.

Slotin was just sixteen when he entered college in Winnipeg, even younger than Daghlian had been. He took up boxing there, perhaps to make himself seem tougher, and adopted an equally macho style of dress: jeans, cowboy boots, a shirt flapping open. After college he toured Spain and later liked to imply (probably falsely) that he'd taken up arms in the Spanish Civil War. Slotin eventually landed a job building cyclotrons in Chicago, then got ensnared in plutonium research. In December 1944 he arrived at Los Alamos and immediately asked to tickle the dragon. The others in the group welcomed him, but even among those daredevils, he earned a reputation as reckless. One time, needing to make some adjustments to a live nuclear reaction taking place in a tank of water, he asked the maintenance crew to shut the equipment down. But it was Friday afternoon, and the crew told him to wait until after the weekend. When they returned Monday morning, the changes had already been made. Had Slotin shut the equipment down himself? No, he'd just stripped off his jeans and cowboy boots and dived into the tank—with the reactor still blazing.

Slotin took pleasure in shocking his family when he revealed his role in building the Hiroshima and Nagasaki bombs. But he soon felt disillusioned with weapons work, and after Daghlian's death he asked to return to Chicago. The military told him he had to arm the bombs for the Bikini tests first. Slotin gave in and agreed, and also agreed to train his replacement in Los Alamos, a man named Alvin Graves.

The training sessions started, and ended, on May 21, 1946, with a demo that Slotin had run maybe forty times before. He first grabbed

Rufus—the same plutonium sphere Daghlian had used—and set it into its cradle. Now he had to set up the neutron reflectors. But rather than use tungsten-carbide bricks, Slotin grabbed a hemispherical shell of beryllium, a more powerful reflector. Beryllium reflected neutrons so well, in fact—it practically played ping-pong with them—that most scientists took extra precautions when using the shells. In particular, they knew that dropping the shell onto the plutonium sphere would cause an immediate chain reaction, so they added a safeguard: they stacked thin wooden shims around the perimeter of the sphere first. They then set the beryllium dome atop the shims and removed them one by one as the experiment progressed. This way they could lower the shell slowly and inch their way toward criticality.

Re-creation of the incident that killed Louis Slotin after the Manhattan Project. (Photo courtesy Los Alamos National Laboratory)

Slotin couldn't be bothered with all that. He simply propped one edge of the shell on the cradle, balanced it at a jaunty angle over Rufus, and used a screwdriver to pump the opposite edge up and down—yo-yoing toward the brink of a chain reaction. Watching him do this once, Enrico Fermi had scolded Slotin that he'd "be dead within a year" if he didn't knock it off. Slotin rolled his eyes.

On that May afternoon, Slotin threw the demo together in two minutes and started jacking the hemisphere up and down. A nearby loudspeaker rumbled with clicks. Graves, peeping over Slotin's shoulder, was impressed. A moment later, however, something happened. The screwdriver slipped, and the beryllium shell came crashing down over the plutonium. All eight men in the room felt a pulse of heat and saw a halo of blue light.

Slotin slapped the hemisphere away with his hand. That hand began tingling a second later, and a metallic taste flooded his mouth. But after he caught his breath, he felt the same eerie serenity that Daghlian had: he knew he'd killed himself, yet for the moment he felt normal. He simply took a step backward and said, "Well, that does it."

At this point Slotin's main concern was the other men in the room, especially Graves. Before he left for the hospital, he tried to estimate the dose of radiation each man had absorbed by using a detector on some objects sitting near Rufus, including a hammer and a Coke bottle. Unfortunately, the detector itself was so contaminated that its readings were useless. As a last resort, Slotin stopped by the desk of a female "computer"—a number cruncher—and asked her to estimate the dose for someone standing where Graves had been. (According to modern estimates, Slotin absorbed the equivalent of 200,000 chest X-rays; Graves absorbed "only" 35,000, mostly because Slotin's body had shielded him.) Cringingly, either because Slotin was dazed or didn't know her, the computer he asked to run these numbers was Graves's wife, Elizabeth, who didn't learn the reason for her calculations until hours later.

Slotin had been working on the same plutonium core as Daghlian. It was the same day of the month as Daghlian's accident, the twenty-first, and was even another Tuesday. And when Slotin arrived at the hospital, doctors put him in the same bed in the same room—and watched him die the same agonizing death, as his body essentially disintegrated. Nine days later they flew Slotin's corpse home to Winnipeg in a casket lined with lead.

The twin deaths of Daghlian and especially Slotin—the "most experienced bomb putter-togetherer" in the world, as he liked to brag—shook the world of nuclear physics and opened a rift within the community. Some physicists already felt guilty over what they'd wrought with the A-bomb. Now, with colleagues dying, their remorse hardened into outright opposition to nuclear weapons. But some scientists reasoned differently. The impending Cold War seemed of far, far greater magnitude to them than the life of any one man. Civilization itself was at stake, and they saw the death of a colleague as a courageous sacrifice. These hawks would soon include the man peering over Slotin's shoulder the day of the accident, Alvin Graves.

After it killed both Slotin and Daghlian, scientists stopped calling that cursed plutonium sphere Rufus and began referring to it as the "demon core." To us today, it might seem tactless, if not crazy, to keep using the demon in more experiments, but plutonium was the most valuable substance on Earth then, and the military wasn't about to throw millions of dollars' worth away. So shortly after Slotin died, the demon core was loaded onto a plane for the South Pacific and ended up inside one of the bombs at Operation Crossroads.

———◦———

Incredibly, the U.S. military originally wanted to vaporize a few of the Galápagos Islands with the Crossroads bombs. They finally settled on Bikini despite the limited space, fragile ecosystem, and (foreshadowing alert!) unpredictable winds there.

The Baker test near the Bikini Atoll, July 25, 1946. (Photo courtesy U.S. Department of Defense)

The Gilda/Able test on July 1 was the fourth atomic bomb blast in history; the fifth occurred three weeks later, in the Baker test on July 25. (The bomb with the demon core would have been the sixth, but the military scrubbed the test and rather unceremoniously decided to melt down the core and recycle the plutonium for use in other bombs.) The Baker bomb, dubbed Helen of Bikini, lacked for much predetonation drama: instead of being dropped from a plane, it exploded ninety feet underwater, to mimic a sneak attack on a fleet. But the postdetonation special effects more than made up for it. Within ten milliseconds, the center of the lagoon lit up like a diamond, refulgent with light, and three million cubic feet of water vaporized into steam. (James Watt would have been agog.) In addition, two million gallons of water went whooshing upward in the biggest fountain the world had ever seen, two thousand feet across and six thousand feet tall. Nine ships sank immediately, killing more pigs and goats and rats. The surviving fleet was drenched in what one study called a "witch's brew" of radioactive water.

Despite the drenching, the navy tried to salvage these remaining ships as well as save any animals on them. How? By sending in thousands of sailors to wash the decks. A good scrubbing with lye, the admirals figured, and a fresh coat of paint should take care of any pesky radioactivity. When this plan failed—the onboard Geiger counters kept chattering away—the admirals were shocked. Could plutonium really withstand whitewash? Eventually, the navy admitted defeat and abandoned or sold for scrap sixty of the ninety ships. (Several remain sunk in the lagoon today, an underwater wonderland for octopus, fish, and scuba divers.) Even more worrisome, after removing the animals, the navy noticed an increasing number of health problems among them: they suffered the same fatigue and weight loss and low blood-cell counts as Pig 311 but never improved.

For the most part the public remained in the dark about these long-term problems. Most reporters left Bikini immediately after the bombs exploded, and they'd generally downplayed the threat of atomic weapons in the dispatches they filed. As mentioned, nuclear bombs had obtained almost mythological powers in people's minds since Hiroshima and Nagasaki. In early 1946 military scientists actually had to put out statements reassuring the public that despite rumors to the contrary, the Bikini tests would not "destroy gravity" or "blow out the bottom of the sea and let all the water run down the hole." So when Gilda and Helen failed to usher in the apocalypse— and in fact left several ships afloat in the lagoon—reporters scoffed. They began to dismiss atomic bombs, mocking them as "distinctly overrated." (One radio jockey claimed to have made a clandestine recording of the Gilda blast, then played a ludicrous chicken peep on the air.) Few did any follow-up reporting or considered the possibility that the real danger might be invisible.

In taking this stand, reporters were also telling the public what the public longed to hear. Call it cowardice, call it human nature, but after several million deaths and forty-four months of inch-by-inch warfare on two continents—and then another year of hysterical

A mushroom cloud cake. (Photo courtesy Library of Congress)

stories about the Bomb on top of that—most people just wanted to get on with their lives, not keep fretting. The anticlimax at Bikini even gave them permission to laugh a little. People threw parties with angel-food cakes shaped like mushroom clouds. Nightclubs boasted of their "anatomic bomb" dancers. And French designers, *naturellement,* released a two-piece swimsuit every bit as compact and dangerous, they winked, as the weapons at Bikini.

This apathy enabled the U.S. government to keep testing bombs—a program that accelerated after the Soviet Union detonated its first nuke in 1949. American test sites ranged almost from sea to shining sea, as far northwest as the Aleutian Islands and as far southeast as Mississippi. Most tests took place in Nevada.

Many of these early trials focused on one question: how well the average American home would stand up to a megaton nuclear blast. Not very well, it turns out. To run these tests scientists erected rows of houses and storefronts in the Nevada desert, which they cheerily named Survival City. (The local roads had grimmer handles: Death Street, Disaster Lane, Doomsday Drive.) Each house was stocked with furniture and food, as well as mannequins doing everyday, nuclear-family things like sleeping in bed, playing with Baby, or entertaining friends with drinks and records.

It was pretty clear from the get-go that nothing much would survive in Survival City. Whenever bombs went off, buildings within a few hundred yards of the zeropoint crumbled into ash. Homes farther afield might remain standing, but most of the dummies inside were burned and splintered; a few mannequin children got decapitated. With determined optimism, however, military scientists announced that things hadn't been so bad. They even dug a few iceboxes out of the rubble and cooked up the frozen strawberries, chicken pot pies, and French fries inside for a focus group, to show how people could expect to live in the Bomb's aftermath. The diners declared the meal delectable.

Health officials, too, fed plenty of tripe to the public. One doctor suggested that far from harming the body, radioactivity actually "stimulates the spermatocytes." He added that "plutonium, next to alcohol, is probably one of the better things in life," and claimed that he used it in his tooth powder. A Harvard psychologist declared that the biggest threat humanity would face after a nuclear attack wasn't, say, millions of deaths or the collapse of civilization, but too much unmarried sex among the survivors.*

All the while, though, unsettling news was circulating about the longer-term and more insidious effects of nuclear weapons. Fast-acting cancers such as leukemia were already decimating Hiroshima and Nagasaki. Manhattan Project scientists, now scattered around the country, began whispering about Slotin and Daghlian over lab benches and at faculty lunches. Even Pig 311, although living high on the you-know-what at the National Zoo, turned out to be sterile. She also ballooned to six hundred pounds and died in 1950 at age four and a half, markedly young for a pig. Maybe radiation wasn't one of the good things in life.

Worse, a scary new word entered the national vocabulary in the early 1950s: fallout. Most radioactive fallout from U.S. weapons testing settled in the Nevada desert, and the government monitored it carefully there. But the government had no national detection program at first, because scientists didn't realize that mushroom clouds could carry fallout high into the atmosphere, where winds dispersed it widely. In fact, the first people to realize just how far fallout spreads were employees of the Eastman Kodak company, who discovered that some of the packing material they used for shipping was radioactive. It was made of recycled corn husks from southwest Indiana, and it was discharging at high enough rates to ruin whole crates of film.

Fallout clouds consisted of several types of particles. The clouds themselves were sometimes tinted red at first from all the nitrogen oxide gases that formed in the heat of nuclear blasts. But the really dangerous stuff was invisible: lone, rogue, radioactive atoms that appeared when plutonium split into fragments, everything from antimony-125 to zirconium-97. Plutonium fission also released neutrons, which stuck fast to otherwise friendly molecules like N_2 and turned them radioactive. These radioactive species then migrated thousands of miles in high-altitude air currents, either settling out on their own or getting snagged by rainstorms and turning unsuspecting cities — Albany, New York, or Minot, North Dakota — into nuclear hotspots. Counterintuitively, driving rains were actually less dangerous, since they produced more runoff and washed radioactivity away; mists, meanwhile, allowed radioactive bits to linger.

The world had never known a threat quite like fallout. Fritz Haber during World War I had also weaponized the air, but after a good stiff breeze, Haber's gases generally couldn't harm you. Fallout could — it lingered for days, months, years. One writer at the time commented about the anguish of staring at every passing cloud and wondering what dangers it might hold. "No weather report since the one given to Noah," he said, "has carried such foreboding for the human race."

More than any other danger, fallout shook people out of their complacency about nuclear weapons. By the early 1960s, radioactive atoms (from both Soviet and American tests) had seeded every last square inch on Earth; even penguins in Antarctica had been exposed. People were especially horrified to learn that fallout hit growing children hardest. One fission product, strontium-90, tended to settle onto breadbasket states in the Midwest, where plants sucked it up into their roots. It then began traveling up the food chain when cows ate contaminated grass. Because strontium sits below calcium on the periodic table, it behaves similarly in chemical reactions. Strontium-90 therefore ended up concentrated in calcium-rich milk — which then got concentrated further in children's bones and teeth when they drank it. One nuclear scientist who had worked at Oak Ridge and then moved to Utah, downwind of Nevada, lamented that his two children had absorbed more radioactivity from a few years out West than he had in eighteen years of fission research.

Even ardent patriots, even hawks who considered the Soviet Union the biggest threat to freedom and apple pie the world had ever seen, weren't exactly pro-putting-radioactivity-into-children's-teeth. Sheer inertia allowed nuclear tests to continue for a spell, but by the late 1950s American citizens began protesting en masse. The activist group SANE ran ads that read "No contamination without representation," and within a year of its founding in 1957, SANE had twenty-five thousand members. Detailed studies of weather patterns soon bolstered their case, since scientists now realized just how quickly pollutants could spread throughout the atmosphere. Pop culture* weighed in as well, with Spiderman and Hulk and Godzilla — each the victim of a nuclear accident — debuting during this era. The various protests culminated in the United States, the Soviet Union, and Great Britain signing a treaty to stop all atmospheric nuclear testing in 1963. (China continued until 1974, France until 1980.) And while this might seem like ancient history — JFK signed the test-ban treaty, after all — we're still dealing with the fallout of that fallout today, in several ways.

First, just for context, I should point out that there's nothing intrinsically evil or even unnatural about radioactivity. If you're one of those people who eats food and drinks water, you already consume radioactivity from natural, non-fallout-related sources — at least four chest X-rays' worth each year. (To be clear: it's not that your food releases X-rays per se; it's that the different types of radioactivity in food do the same amount of damage to your tissue* as four X-rays would.) There's radioactivity in Brazil nuts, in coffee, in red meat. Bananas contain enough radioactive potassium-40 that large shipments sometimes set off radiation detectors at seaports. Nuclear scientists have even defined a fanciful measure of everyday radioactivity called the BED, or banana equivalent dose.

(Incidentally, all that potassium-40 — both in bananas and in Earth's crust generally — gradually decays into argon and seeps into the air. That explains why this random noble gas is so common in the atmosphere, because radioactive potassium is constantly manufacturing it.)

Speaking of air, if you happen to be a breather, you inhale twenty more chest X-rays of radioactivity each year, mostly from radon. There are cosmic rays to fret over as well, streams of subatomic particles that originate in deep space and pelt our planet in unimaginable numbers (up to ten thousand per square meter per second in some spots). The atmosphere filters out most cosmic rays, but because air is thinner at higher altitudes, you expose yourself to more of this nuclear hail every time you take a cross-country flight or visit Denver.

I could go on. Smoke detectors release alpha particles, old television sets release X-rays. Kitty litter oozes uranium, as do glossy magazines and trendy granite countertops. And given that you're not keeling over every time you sift Mr. Whiskers's box, you can probably guess that there's no need to worry about everyday radio-activity. You'd need to eat twenty million bananas to induce radiation poisoning, eighty million to guarantee death. All that talk about dozens of chest X-rays sounds scary, but remember that this damage gets spread over a whole year, which gives your cells time to recuperate and repair any damage.

So, how much lingering radioactivity do we inhale each year from atomic-weapons testing? Cue the sad trombone: about a tenth of an X-ray's worth. This will shorten the life-span of the average person by 1.2 minutes. Statistically, four puffs on a cigarette do more harm.

Before you snort *big deal,* though, consider this. Natural cosmic rays often produce radioactive carbon-14 in the air. Carbon-14 then latches onto oxygen to form radioactive CO_2, which in turn gets absorbed by plants and starts to work its way up the food chain. You have around a million billion atoms of this radioactive carbon inside you right now, and it's constantly unloading on your cells. While not overwhelming, these atoms do damage and mutate DNA, and cancer can result.

With that in mind, now consider the fact that atmospheric bomb testing almost doubled the amount of carbon-14 in the air between 1950 and 1963, adding more than a trillion pounds. That concentration still hasn't dropped back to normal, either, and won't for thousands of years. As a result, you have a higher chance of getting cancer than you would have otherwise, even if you were born after the test-ban treaty. How much higher? Depending on the assumptions you make, the extra carbon-14 was inducing between a hundred thousand and a million extra DNA mutations per person per day as late as 1990. To be sure, the body's normal mutation rate dwarfs that by several orders of magnitude, and most of the damage will be repaired quickly. But not all. Across the world, scientists estimate that we will suffer several million extra cases of carbon-14-related cancer thanks to nukes, including two million extra cancer deaths, when all is said and done — with most of them still to come. (Remember, too, that this is just carbon-14; other radioactive species are still floating around as well, flitting into and out of your cells.) All this helps put that statistic about 1.2 minutes into perspective. It's accurate, but it's an *average*. Most people will lose zero minutes. But millions of people will lose far more than eighty-some seconds.

Along those same lines, it's somewhat misleading to compare absorbing X amount of radioactivity in dribs and drabs over a whole year with absorbing X amount of radioactivity in one big gulp of fallout exposure, since big gulps do more damage. Sailors who worked on the Bikini bomb tests died, on average, three months earlier than people in control groups, which isn't so trifling. People born during the baby boom of 1946–1964 face grave risks, too, since they were children during the height of bomb testing.

As an example, consider my mother, who had several demographic strikes against her. She grew up in the 1950s, and she grew up in rural Iowa, which got peppered with far higher concentrations of fallout than population centers along the coasts. (In general, kids in Utah, Montana, and Idaho had it worst, due to prevailing winds from Nevada. In the continental United States, kids in Los Angeles had it best, since winds from Nevada rarely blew that way.) Most of my mom's exposure would have come from drinking milk contaminated with strontium-90 and, worse, iodine-131, another common fission product. Like strontium, iodine is absorbed by plants and concentrated in cow's milk; breast milk also contains appreciable amounts, as do cottage cheese, eggs, and leafy vegetables. And while strontium-90 at least gets distributed throughout the body, iodine-131 concentrates in the thyroid gland, bombarding that tongue-sized organ with radioactive particles. Iodine-131 does decay quickly, with a half-life of eight days. But it took only a few days for the reddish cloud of fallout to drift from Nevada to Iowa. My mother's family compounded the problem by sometimes drinking fresh milk from the cow in their cousins' backyard instead of store-bought milk, which sits on the shelf for a few days and gives the iodine-131 time to deplete. Based on numbers provided by the U.S. government, her thyroid gland probably absorbed five times more iodine-131 than that of your typical American then, and a thousand times more than I ever will (touch wood). Some of her peers out West absorbed at least fifteen times the average, possibly more.

So what are the medical consequences of all this? Geneticists feared a rash of mutant babies among the "downwinders" in the 1950s, but that never came to pass. For a child to inherit a mutation, the DNA in Mom's eggs and/or Dad's sperm has to be scrambled, and strontium-90 and iodine-131 don't congregate near the gonads, thankfully. The real worry, again, is cancer. Based on pre-1950 rates, doctors have calculated that people who grew up in the 1950s in the United States should come down with something like 400,000 cases of thyroid cancer over their lifetimes. Radioactive iodine-131 will probably boost that by 50,000 cases. Rates of other cancers will climb as well.

Let's not forget that people beyond the United States also suffered, especially the poor Bikini Islanders. The U.S. military had evicted all 167 natives in 1946 on the promise that they could return within a few years. Unfortunately, Gilda and especially Helen poisoned the fish in the lagoon on which they depended for food. (Biologists recovered one fish in 1946 that was radioactive enough to take a "self X-ray": they just plopped it onto some photographic paper, and an outline of it appeared a few hours later.) After their eviction, the Bikinians eked out a living on a nearby atoll that lacked fresh water and fishing holes, leaving them half starved and fully dependent on the U.S. military for supplies. Although a few Americans had compared the initial move to the Israelites being led into paradise, a Bikini chieftain named Judah pointed out, much more aptly, that their situation now called to mind the Israelites wandering in the desert. They were nuclear nomads.

They still might have been able to return to Bikini a few years later if the U.S. hadn't started testing bigger, badder nukes called Supers there. Like many of Alfred Nobel's experiments, the Supers were two-stage devices. First came a plutonium fission blast. The X-rays from this blast then excited a nearby cache of hydrogen atoms, which started fusing into helium. This same basic fusion reaction powers the sun, releasing gobs upon gobs of energy in tiny fractions of a second. On March 1, 1954, the U.S. military unleashed this heavenly hell on Bikini, in the infamous Castle Bravo test.

The scientific director for Bravo was none other than Alvin Graves, the man peeking over Louis Slotin's shoulder the day of the second demon core accident. Graves had barely pulled through the ordeal, and he'd endured a grueling recovery: after returning home from the hospital, he slept sixteen hours most days and his sperm count dropped to zero for a while. Nevertheless, both Graves and his wife (who'd blithely run the absorption calculations on him) ardently supported nuclear weapons. Perhaps they had to, to justify all they'd suffered.

By 1954 Graves had fathered two children, and despite his eyes being clouded with cataracts, he flew to Bikini to direct the Bravo test.* Things got bollixed from the start. For one, the bomb released far more energy than expected, the equivalent of three billion pounds of TNT—650 times Helen's yield. Worse, despite some dodgy weather that morning, Graves gave the go-ahead to detonate anyway, and a wind kicked up at the exact wrong moment, scooping up the cloud of fallout and dumping it onto a cluster of islands east of Bikini.

The tribes on those islands saw a gigantic red fireball explode onto the horizon at 6:45 a.m.—a sunrise in the wrong direction. A few hours later a white, salty ash started wafting down. Within a day, islands 150 miles distant had received flurries of this "Bikini snow," and children there (who'd never seen real snow and didn't know any better) began to play with and eat it. In addition, twenty-three Japanese tuna fishermen, on a not-so-fortunate boat named *Lucky Dragon,* ran into the blizzard as well. Within a few days, fishermen and islanders alike were complaining of headaches, nausea, fatigue, and rashes, and their hair had started falling out. Children on the closest atolls absorbed more than one hundred times the fallout that the worst American cases had—equivalent to around ten thousand chest X-rays at once. On one island, fifteen of nineteen children developed thyroid tumors before age twenty-one. Two lost thyroid function entirely and stopped growing; one died of leukemia. All that suffering because of a rogue east wind.

The truth about the weather near Bikini that morning remains mushroom-clouded even today. Some insist that a fanatical Graves ordered the test to proceed despite knowing that the wind would dump the fallout on the other islands' laps. Others say that a crew of military meteorologists bungled their forecasts, still others that the combination of a bigger-than-expected bomb and swirling winds made the job of forecasting nigh impossible. Perhaps some combination of all three. Still, I suspect the third explanation is closest to the truth. Knowledge about the behavior of weather fronts was pretty limited in the 1950s and contributed to the early nonchalance about fallout. As we'll see in the next chapter, our atmosphere is one of the most complicated physical systems in existence, and even in our age of supercomputers and superbombs, truly accurate weather forecasts remain elusive. This shouldn't be so. Weather fronts are just pockets of warm or cold gases flowing across the surface of the Earth, after all.

Interlude: Albert Einstein and the People's Fridge

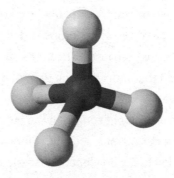

Dichlorodifluoromethane (CCl_2F_2) — currently 0.00054 parts per million in the air; you inhale seven trillion molecules every time you breathe

Many people know that work on nuclear weapons enabled the development of the first electronic computers. But it's no less true that the humble refrigerator, in a roundabout way, enabled the development of the first atom bomb.

While reading the newspaper one morning in 1926, Albert Einstein nearly choked on his eggs.* An entire family in Berlin, including several children, had suffocated a few nights before when a seal on their refrigerator broke and toxic gas flooded their apartment. Anguished, the forty-seven-year-old physicist called up a young friend of his, the inventor and scientist Leo Szilard. "There must be a better way," Einstein pleaded.

Szilard ("Sil-ard"), a stocky man of twenty-eight, had first impressed Einstein six years earlier by proving him wrong on a certain scientific point. (That didn't happen often.) Szilard also had a knack for turning esoteric ideas into useful gadgets. In later years he became a sort of Thomas Alva Edison of high-energy physics, sketching out the first electron microscope and particle accelerator; he and Einstein had bonded in part over their love of such mechanical devices. (Although a theorist and somewhat flighty, Einstein came from a family of tinkerers—his uncle Jakob and father Hermann had invented new types of arc lamps and electricity meters—and he'd worked in the Swiss patent office for seven years.) So when Einstein called Szilard that morning, the two men agreed to collaborate and build a better, safer refrigerator.

This wasn't as odd as it might sound: in the previous half century, refrigeration had become serious science. The study of thermodynamics and heat had led to the concept of absolute zero—the coldest possible temperature—and several labs around the world were racing to reach the bottom of the thermometer. Some of the best science revolved around attempts to liquefy certain gases: nitrogen, oxygen, hydrogen, methane, carbon monoxide, and nitric oxide. Throughout the 1800s this sextet—the so-called permanent gases—had resisted all efforts to liquefy them (hence the name). This stubbornness had led some scientists to declare that these six gases could never be liquefied, that they somehow stood apart from the rest of matter. Other scientists said baloney—that powerful new cooling methods would eventually condense them. In particular, the latter group pinned their hopes on a clever, cyclical cooling process that involved removing heat from substances in several stages.

Stage one involved filling a chamber with a gas that was easy to liquefy. Call it A. Scientists first compressed A with a piston, then cooled down the compression chamber with an external jacket of cold water. As soon as A had chilled down, a valve opened. This dropped the pressure on A and allowed it to expand into a larger volume. The key point is that expanding into a larger volume takes energy, takes work. (It's similar to how a litter of puppies, if locked in a broom closet, would suddenly expend a lot more energy if you opened the door and let them run free inside the house.) And in this situation, the only energy A can draw on to expand and spread is its own internal store of heat energy. But depleting its internal store of heat energy inevitably cooled A down even more, and it eventually condensed into a liquid at around –100°F.

Now came the clever part. The next stage involved a chamber of gas B, which was tougher to liquefy. Scientists once again compressed B with a piston to start. But for the cooling jacket this time, instead of cold water they ran liquid A through the jacket. This dropped gas B's temperature to –100°F. Opening a valve then caused B to expand, which forced B to deplete its internal store of heat energy. Its temperature plunged to around –180°F, whereupon it also liquefied.

Liquid B could now be used in another cooling jacket to liquefy a more stubborn gas, C, and so on alphabetically. This bootstrapping process finally reached temperatures so low (circa –420°F) that not even "permanent" gases could resist, and all six were eventually liquefied.* Especially beautiful was liquid oxygen, which glowed faintly blue, like liquid sky.

Gas refrigeration remained a mere curiosity, however, until the Guinness Brewing company invested in the technology around 1895. Before this, breweries generally brewed beer only in the winter and stored it. (*Lager* means "storage" in German.) Refrigerators let Guinness make beer year-round, thank goodness. As a knock-on technology, the rest of the world got commercial refrigerators, like the one in your home right now. All modern fridges rely on the same general principles of gaseous cooling.

If you tore out the inner panels on your fridge, you'd see a series of tubes. Inside the tubes you'd find a liquid (call it Z) with a low boiling point. As the casseroles and other leftovers inside your fridge emit heat, Z absorbs the heat through the fridge walls and warms to a boil. The resulting gaseous Z then floats away through other tubes, carrying the heat with it.

Next, Z enters a compression chamber, which compacts the gas with a piston. (The motor that runs the compressor causes the characteristic hum of refrigerators.) The compressor now pushes warm gas Z through still more tubes behind the fridge, which allows Z to jettison heat to the outside world. At this point the gas has successfully removed heat from inside the unit and dumped it out back. And after Z dumps enough heat, it condenses back into liquid. Now Z passes through an expansion device that lowers its pressure, cools it further, and completes the cycle. Liquid Z reenters the tubes inside the fridge panels, reboils, and resumes sucking out heat.

Now, one detail here might sound suspicious. You're boiling a liquid (Z), so shouldn't everything heat up? Not quite. The *liquid* heats up, yes. But in an enclosed unit like a refrigerator, the liquid can warm itself up only by stealing heat from your casserole: warming the one necessarily cools the other. And the boiling is indeed crucial. Remember James Watt's old bête noire, latent heat? This principle says that liquids changing into gases absorb ridiculous amounts of energy. In Watt's engines this was a bug, but fridges make it a benefit: absorbing heat and whisking it away is exactly what refrigerators aim to do, and nothing does that better than liquids changing into gases. (This same general process explains why liquid sweat, when it evaporates, cools you on a summer day.)

By the 1920s gas-compression refrigerators had replaced iceboxes all across Europe and North America. There was only one problem. All three gases commonly used as coolants then—ammonia, methyl chloride, and sulfur dioxide—were toxic and occasionally killed whole families. (Methyl chloride sometimes exploded, too, just for fun.) Hence Einstein's vow to find "a better way." He knew the weak point in home refrigerators was the compressor, whose seals often cracked under pressure. So he and Szilard designed a fridge without a compressor, a so-called absorption fridge.

In the simplest type of absorption fridge you start with two liquids mixed together in a chamber, the absorbent and the refrigerant. (Mark those names.) The key to the design is that, at low temperatures, these substances mix readily. But if you raise the temperature — usually by warming the chamber with a small methane flame — the refrigerant boils out as gas, leaving the absorbent behind.

The refrigerant gas now goes on a long and tortuous journey. It first flows into tubes behind the fridge and dumps the heat it absorbed from the flame; this step simultaneously cools the refrigerant back into liquid. This liquid flows via gravity into the panels inside the fridge, where it sucks the heat out of yet another casserole. Absorbing this heat causes the liquid to reboil, and the resulting gas whisks the latent heat away, removing it from the unit's interior. (In some designs the gas then heads to still more tubes behind the fridge, to jettison heat one last time.)

Meanwhile, back in the original chamber, the methane flame has switched off, allowing the absorbent there to cool down. A jacket of cold water then cools the absorbent further. The absorbent cools so much, in fact, that when the refrigerant gas finally wends its way back into the chamber, the absorbent condenses it into liquid again and reabsorbs it. You therefore end up back where you started, with a mix of two liquids that you can separate with a flame. Overall, absorption fridges and regular fridges cool things down the same way, by boiling gases. But they use a different process to recycle the refrigerant.

Again, though, this probably sounds like cheating: a flame can cool my beer? But that's the magic of gases. Really, the flame here isn't so much adding heat as doing physical work—separating the refrigerant from the absorbent by turning the refrigerant into gas. And once you have a free gas in the system, you have oodles of options. Indeed, the art of refrigeration consists of manipulating gases to absorb heat energy here, carry it there, and dump it somewhere else. Hearkening back to Thomas Savery, you could call the Einstein-Szilard refrigerator an engine for freezing water by fire.

Leo Szilard (right), inventor of the nuclear chain reaction, collaborated with Albert Einstein on the invention of several types of refrigerators. (Courtesy of Los Alamos National Laboratory)

The Einstein-Szilard fridge actually used three liquids and gases, not two, making it a tad more complicated than the scheme above. But their design did have several advantages over regular fridges. With no motor, it made no noise and rarely broke down. It also used no electricity (just methane), and it avoided the seals that all too often broke and leaked toxic gas.

In looking back on this episode, some historians have assumed that Einstein merely offered advice on the patent applications or used his celebrity to lure investors, leaving the real work to Szilard. In truth Einstein labored over the project, and the duo ended up receiving dozens of patents in six countries on different fridge components. (An American patent attorney reviewing applications did a double take, as well he might, when he noticed Einstein's signature.) The duo ended up selling several patents and collecting a nice check for $750 (around $10,000 today); they subsequently opened a joint checking account, like a married couple. Szilard collected an additional $3,000 per year in consulting fees.

Like any married couple, though, they clashed sometimes. Szilard had an engineer's appetite for complexity and kept adding new valves and cooling lines to the fridge. Einstein, meanwhile, longed for simplicity and elegance — no less in his home appliances than in his physics. (He would have hated working with James Watt.) The need for simplicity eventually drove Einstein and Szilard to invent two other cooling units, each of which worked on a different physical principle. In one they replaced the piston in a standard fridge with molten sodium, which magnets pumped up and down to compress gases. The other device used water pressure from a kitchen faucet to power a small vacuum pump; the pump then cooled things by evaporating methanol. Einstein called the latter device *Der Volks-Kühlschrank,* the people's fridge.

In the end, sadly, none of the three Einstein-Szilard fridges ever made it into anyone's home. Not surprisingly, the molten-sodium pump proved a wee bit impractical for your average kitchen (though it later found use in nuclear power plants). The faucet cooler failed because German apartment buildings had lousy water pressure, which hindered the vacuum pump. And absorption fridges simply burned too much fuel to compete with compression fridges; the Einstein-Szilard design seemed like a Newcomen engine in comparison.

Even the biggest objection to conventional fridges, the lethal gases, became moot in 1930 with the debut of a new and nontoxic cooling gas, Freon (CF_2Cl_2). Within a decade, virtually all home units had switched to this chlorofluorocarbon,* and the Einstein-Szilard fridge was rendered a historical relic. Of course, Freon did have one pesky drawback. When old refrigerators went to the junkyard, the Freon leaked out and climbed into the stratosphere. There, ultraviolet light cleaved the chlorine atoms off, creating free radicals that chewed through ozone molecules with sickening efficiency: each chlorine radical can destroy 100,000 O_3 molecules over its lifetime. This destruction eventually opened up a hole in the ozone layer that still exists and that won't recover for decades, if ever. Humanity might have saved itself a lot of trouble in the long run by investing in the Einstein-Szilard approach to cooling water with fire.

(Incidentally, the chemist who invented Freon, Thomas Midgley, also developed the first leaded gasoline in 1921. He did so to help curb engine knock, but the lead in the gasoline also polluted the atmosphere and damaged the brains of growing children. In less than a decade, then, this one man invented arguably the two worst industrial products of the twentieth century. Midgley never lived to see what he had wrought, however. After coming down with polio in 1940 and losing the use of his legs, he invented a system of ropes and pulleys to move himself from bed to wheelchair. He got tangled up in them one morning in 1944 and strangled himself.)

So was the Einstein-Szilard fridge a waste of these men's time and talent? Not entirely. Einstein found the work a refreshing break from his futile search for a Theory of Everything. With two families to support and a crumbling German economy, Einstein also enjoyed the extra cash. Szilard needed the money even more, especially after he fled Nazi Germany for London in 1933. (He was part Jewish.) He spent the next few years living off his fridge proceeds, and he used his sudden freedom to take long walks and ponder what the next big thing in physics might be. The answer came to him one afternoon in September 1933, as he stepped off a curb near the British Museum. He'd been hearing about some experiments involving the release of subatomic particles called neutrons. He started wondering what would happen if, say, a uranium atom split and released multiple neutrons. Other nearby uranium atoms might absorb them, become unstable, and release neutrons themselves when they split. These secondary neutrons would destabilize more atoms, which would release tertiary neutrons, and so on. Each atom that split would also — according to his patent partner's famous equation, $E = mc^2$ — release energy in an ever-growing cascade...

By the time he crossed the street, Szilard had worked out the principle behind the first nuclear chain reaction. And unlike his clever fridges, this invention became all too pervasive in the turbulent decades to follow — decades that would shatter not only the public's belief in benevolent science, but scientists' belief in a neat, tidy, predictable universe.

CHAPTER EIGHT

Weather Wars

Silver Iodide (AgI) — currently zero parts per million in the air (unless someone is seeding clouds above you)

Chemist Irving Langmuir had already won a Nobel Prize, but he'd never screamed in delight during an experiment before. It was November 13, 1946. He was standing in a control tower at the Schenectady, New York, airport, watching a small prop plane go buzzing overhead. Fourteen thousand feet above him, his assistant was leaning out the plane's window, tossing pellets of dry ice into a cloud. Seconds later, the cloud "began to writhe as if in torment," one witness recalled. Within five minutes, the cloud had disappeared, transformed into rain.

None of this rain actually reached the ground, mind you — it evaporated before that. Nevertheless, Langmuir practically began running around in circles down below. "This is history!" he cried. Even before the plane landed, he raced off to telephone a reporter. Mankind, he shouted into the receiver, had finally learned to control the weather.

Had anyone else made this pronouncement, the reporter likely would have hung up. But in his day Langmuir was as famous as Albert Einstein, his opinions just as highly regarded. And although Langmuir was a chemist by training, this foray into meteorology didn't strike anyone back then as foolhardy; outsiders used to make forays into meteorology all the time, in fact. Pneumatic chemists studied the weather intently to understand how gases like air behaved. Astronomers took assiduous notes on the weather in order to predict when the skies would clear for their telescopes. So did doctors, on the theory that bad air caused disease. Robert Hooke, John Dalton, James Watt, Lord Rayleigh—the list of sometime meteorologists goes on and on. Charles Darwin, during the voyage of the *Beagle,* became the first scientist to study El Niño, and the *Beagle*'s captain, Robert Fitz-Roy,* published the first weather forecast in history, in *The Times* of London in 1861.

Still, Langmuir pushed far beyond his predecessors in his hopes and ambitions for meteorology: he wanted not only to understand the weather but control it. To be sure, scientists dabbling in meteorology had always had a streak of irrational confidence. Despite centuries of dashed hopes, they never lost faith that they were *this close* to understanding how the weather worked. Just wait until we get our new barometers, our new weather stations, our new computers—then we'll predict the weather perfectly. Meteorology is the Bad News Bears of science, perpetually just one year away.

Even among these Panglosses, though, Langmuir's optimism stands out. He came to meteorology during something of a midlife scientific crisis, and his charisma and glittering credentials persuaded hundreds of colleagues to join his quest. Their work ultimately proved a bust, little more than a modern version of a primitive rain dance, but while it lasted, they put on a heck of a show.

———<◦>———

The first prominent advocate for weather control sprang from the land of can-do, the American frontier. James "the Storm King" Espy, a Kentucky-bred meteorologist, noticed in the 1830s that Indian bonfires sometimes produced rain; on trips to the big city, he saw that the smoke pouring out of factory chimneys seemed to attract rain clouds, too. So on the iron-clad law that correlation always implies causation, Espy declared that smoke must make rain fall, and he began promoting a scheme to regulate rainfall across the eastern United States. All the government had to do was set a massive forest fire on the Appalachian Mountains every Sunday afternoon. Soon our weather would be as regular as the tides.

To be fair, Espy pioneered several nonkooky theories as well, most importantly on the formation of clouds. According to him, clouds form when pockets of warm air ascend into the cooler upper atmosphere and the water vapor inside them condenses. Not only was this theory largely correct, it foreshadowed one of the most important laws of modern meteorology, that water vapor drives most changes in weather. Unlike with most gases in the atmosphere, the concentration of water vapor can vary by several orders of magnitude depending on local conditions, from virtually zero in deserts to a few percent in rain forests. Moreover, unlike other major constituents of air—oxygen, nitrogen, argon, all of which remain gaseous until several hundred degrees below zero—water shifts readily between liquid and vapor within the range of normal, everyday temperatures found on Earth. As a result, water is constantly condensing and evaporating in different places. And because water is constantly condensing and evaporating, it's constantly sucking latent heat out of and dumping latent heat into the air around it. This heat flow causes changes in temperature and pressure, which in turn produce distinct zones of air called fronts. These high- and low-pressure fronts then collide and create weather patterns, by inducing winds to blow or erupting into storms. (The term "front" was actually coined during World War I, inspired by the clashing of armies.) All this chaos from a little bit of water.

The study of water also lured Irving Langmuir into weather research. In his day job, Langmuir studied surface chemistry at General Electric Labs in upstate New York. In contrast to most corporate labs, at GE he had a free hand to research whatever he fancied, and during World War II, he began studying the buildup of ice on airplane wings. This led to a series of field studies at nearby Mount Washington, in New Hampshire. The mountain was famed among weather buffs for its winds: until the 1990s it held the world record for the fastest wind speed ever measured on Earth, 231 miles an hour, in 1934. But Langmuir was more interested in the odd humidity on the mountain: it often produced mists of "supercooled" water that, despite registering far below 32°F, refused to freeze into ice. This Schrödinger's cat–like indeterminacy—how could water not freeze below its freezing point? intrigued Langmuir, and he wanted to know more.

To help with the work, he engaged an assistant named Vincent Schaefer. Whereas Langmuir had several advanced degrees and had studied science in Paris and Germany, Schaefer had dropped out of high school to work at GE and help his parents pay the bills. He started there as a machinist and model maker, à la James Watt, but found the job tiresome and started exploring other options through a correspondence school. (At one point, he seriously considered becoming a tree surgeon.) Meeting Langmuir awakened an interest in natural science, and not long afterward he invented a machine to preserve impressions of snowflakes. Impressed, Langmuir recruited Schaefer in 1946 to help him study supercooled water.

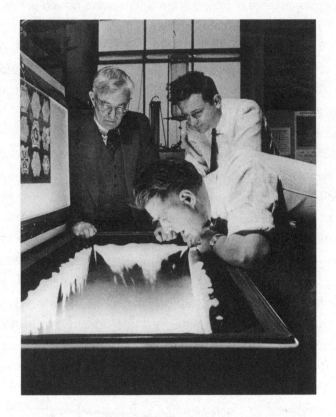

Irving Langmuir (left) and Bernard Vonnegut (right) watch as Vincent Schaefer huffs into a freezer to create ice crystals, a precursor step to making artificial rain. (Photo courtesy miSci-Museum of Innovation and Science)

Schaefer started his experiments by commandeering a $240 open-top GE freezer ($3,000 today). He lined it with black velvet so he could see any ice crystals that formed, then huffed into the cold air to introduce moisture, which became supercooled. Yet week after week, no matter how he varied the conditions in the freezer, the water in his breath never condensed into ice.

One sweltering July day, when the freezer was struggling to keep cold, Schaefer popped over to the lab next door and borrowed a block of dry ice (frozen CO_2) to tuck into one corner. It changed everything. The instant he lowered the cube into the freezer, millions of ice crystals began twinkling in the mist. They then wafted down onto the black velvet, glittering like microscopic diamonds. Schaefer thought at first that the dry ice had induced a chemical change in the mist, but further experiments ruled that out. Rather, the temperature of the dry ice seemed to be the key. Whereas the temperature in the GE freezer bottomed out at –9°F, the frozen CO_2 was below –100°F. When exposed to such brutal, unnatural cold, even supercooled water said uncle and formed ice.

The discovery got Langmuir thinking. Scientists at the time knew that clouds in the sky were basically loose bags of supercooled water. They also knew that most rain actually begins falling from the sky as ice crystals, which melt on their way down. Langmuir reasoned that if he peppered clouds with dry ice, perhaps he could shock the supercooled water and create rain artificially. This led to his renting a plane that November and sending Schaefer up with six pounds of dry-ice pellets, just to see what happened. Twenty minutes later Langmuir was roaring about making history.

Never one to take it slow, he scheduled a doozy of a second test for December 20—and quickly discovered that his freedom at GE had limits after all. As before, Schaefer seeded a cloud that Friday afternoon from a plane, and the cloud once again began to squirm and wriggle. Nothing else happened immediately, so the team went home. But that night eight inches of snow buried upstate New York. Cars went skidding off roads; highways were clogged for hundreds of miles; businesses lost millions in pre-Christmas shopping revenue. Langmuir couldn't have been happier. True, he admitted, the Weather Bureau had predicted the storm; moreover, the cloud they'd seeded had looked "ripe" and might have snowed on its own. But he took full credit anyway—the first purposeful storm in human history.

GE's lawyers, meanwhile, had a collective aneurysm: the potential liability was staggering. They made Langmuir issue a statement denying that the cloud seeding had caused the snowstorm, then forbade him from doing any more fieldwork in GE's name.

Naturally, the injunction disappointed Langmuir, but he was not a man to sulk. His team continued to toil in the lab and—lawyers be damned—he soon sketched out an idea so revolutionary that he abandoned every other project on his slate to pursue it. It promised not only to improve rainmaking but to give Langmuir the superhuman power to control hurricanes.

The idea built on James Espy's general theory of cloud formation. Espy said that clouds form when pockets of warm, less dense air rise into the sky. At some point the water vapor in them cools and condenses into droplets of liquid water. We on the ground see these collections of droplets as clouds, and for many years meteorologists assumed that rain followed automatically, whenever these droplets got around to falling. Turns out it's not so simple. Most droplets that form within clouds *don't* automatically sprinkle down as rain. They're too small. As early balloonists knew, air provides a buoyant upward force on anything suspended within it, water droplets included. And when droplets form at high elevations, most of them are so tiny—one ten-millionth of a gram—that gravity can't overcome the buoyant force and drag them down. Gravity keeps losing this battle, in fact, unless the droplets grow a million times larger, to a tenth of a gram. Clearly, then, for actual rain to fall, a million teeny droplets have to glom together into a larger unit. Otherwise they just sit there.

The obvious question, of course, is what makes the tiny droplets glom together. Intuitively, you might think that droplets simply collide at random and stick to one another. This process isn't very efficient, though, and drops that form this way rarely grow large enough to precipitate out. A better way involves "seeds," solid surfaces for water droplets to latch onto. For various reasons, once a few droplets latch onto a seed, many more follow in quick succession. As a result, droplets can finally grow heavy enough to fall out of clouds as precipitation. If you want to transform a cloud into rain, seeds are vital.

Ice crystals make the best seeds: specks of ice within clouds tend to vacuum up every other water droplet in the vicinity. (This explains why dry ice caused such a ruckus within clouds, because the dry ice converted the supercooled water there into ice crystals.) Foreign particles like dust form great seeds as well. Even airborne bacteria can serve as seeds. Many snowflakes in fact start off life as icy sepulchers for bacteria. (Think about that the next time you're catching snowflakes on your tongue.)

Now, dust and bacteria are fine seeds as far as they go. But Langmuir realized that artificial chemicals might do an even better job. Specifically, he wanted a chemical whose molecular structure mimicked ice — something whose shape would fool supercooled water droplets into latching on. So he ordered another GE assistant, Bernard Vonnegut, to find such a chemical. Over the next few weeks Vonnegut spent many, many stimulating hours in the library paging through crystallography textbooks. He eventually found three candidates, including silver iodide. Silver iodide doesn't resemble water at all chemically — it's a pale yellow powder — but it does form hexagonal crystals like ice does. And tests in that workhorse GE freezer showed that it did indeed fool supercooled water, causing a veritable chain reaction of ice formation. If this worked in real clouds, it would give them a literal silver lining — and allow Langmuir to manipulate them at will.

(This chain reaction with ice led to an interesting side note. Bernard Vonnegut's brother worked in the GE publicity department, and he heard a story one day about how science-fiction writer H. G. Wells had once visited Langmuir's lab. Langmuir took the opportunity to pitch Wells a novel about a scientist who invents a special form of ice that crystallizes at room temperature, forcing all the world's oceans to solidify in a chain reaction. Wells said no thanks. But the idea captivated Bernard's kid brother, and Kurt Vonnegut later spun it into the novel *Cat's Cradle*.)

With GE still skittish about liability, Langmuir had to seek out another partner to sponsor new field tests. He finally found his mark with the U.S. military, which teamed up with him for Project Cirrus in 1947. Project Cirrus had several aims, including relieving droughts. But above all its sponsors wanted to neuter hurricanes, nature's most destructive storms. This pursuit would show Langmuir at his best and worst.

The plan to neuter hurricanes involved a series of observations and deductions, with each step built upon the last. The observations began with the structure of hurricanes, which are basically swirling windstorms centered around an eye. Per the cliché, the eye is calm, but the boundary between the eye and the rest of the storm—called the eyewall—is the most destructive part, with wind speeds often topping 150 miles per hour. That's awfully fast, and you'd think a vortex that violent would tear itself apart. After all, anything that spins should "feel" a centrifugal force pulling it outward, the same force that makes you fly off a merry-go-round if you aren't holding on tight. Hurricanes are no exception: this outward-tugging force threatens to wrench them apart. There's another force to consider, however, one based on pressure differences. You see, the eyewall consists of air at high pressure, the eye of air at low pressure. And because air always tries to flow from high pressure to low pressure, this second force acts *inward,* countering the centrifugal force and keeping the hurricane compact. In other words, despite the violent winds inside hurricanes, they remain stable because these two forces balance.

For Project Cirrus, then, the task was simple: disrupt this balance. The easiest way to do this, Langmuir saw, was by changing the temperature of hurricanes and letting the magic of the ideal gas law* do the rest. His plan was this. Pilots in sturdy airplanes would plunge into the eyewall and seed it with dry ice or silver iodide. Per the discussion above, these chemicals would act as seeds and force any supercooled water there to transform into ice crystals. In a cloud, of course, this transformation would lead to rain. But within hurricanes, Langmuir was actually focused on something else, on a side effect of ice formation—the release of heat.

Transforming water into ice always releases heat, actually. Once again that might sound backward—ice is cold, not warm!—but it makes sense if you break the process down. Think about ice melting in your hand. Why does it melt? Because it's absorbing heat from its surroundings. That's why it feels cold, because it's sucking heat out of your fingers. Now consider the reverse process, water congealing into ice. If melting ice absorbs heat, then by the symmetry of physics, water congealing into ice must, must, must release heat. There's no way around it. That's in fact why ice forms: before they settle into ice, water molecules have to slow down, and they can't slow down unless they jettison heat to their surroundings.

This relates to hurricanes as follows. As ice forms in the eyewall, the surrounding air absorbs the heat that gets released. Absorbing heat should make this air expand according to the gas laws. Once it expands, the pressure should drop, because the air molecules are farther apart. This in turn should reduce the pressure *difference* between the eye and eyewall. As a result the inward-facing force that knits the storm together should decrease. The outward centrifugal force should therefore get the upper hand and widen the eye.

Almost there. Widening the eye wouldn't just make a hurricane go poof and disappear. Storms are too big for that. But widening the eye would decrease the wind speed. That's because the speed of a spinning object depends on its width. That might sound a little obscure, but we've all seen this happen in figure skating. Whenever (insert your favorite skater here) spins with his arms tucked in (small radius), he spins fast. When he widens his arms, he slows down. Same with hurricanes: when the eye widens, they slow down. And widening the eye is critical, because the destructive force of a hurricane depends on the square of the wind speed. So cutting the speed by 10 percent would lower the destruction by almost 20 percent. Cutting the wind speed by a quarter would cut the damage by over 40 percent.

To summarize this chain of deductions, Langmuir proposed that seeding hurricanes would create ice and release latent heat; this in turn would widen the eye and reduce the destructive power of the storm. The idea just needed testing—and here's where the trouble started.

On October 13, 1947, a mild hurricane named King sliced through Miami and began drifting northeast, out into the Atlantic Ocean. Because King seemed to be dying anyway, Cirrus officials decided to seed it the next day. A B-17 puttered out to meet it and scattered 180 pounds of dry-ice pellets into the eyewall. Everyone sat back and waited for the eye to widen and for King to collapse. Instead, the storm grew stronger, fiercer. To everyone's horror, it then pivoted—taking an impossible 135° turn—and began racing back toward shore. A few hours later the now-runaway storm plowed into Savannah, Georgia, causing $3 million in damage ($32 million today) and killing one person.

Cirrus scientists hoped like hell that no one had noticed, but a meteorologist in Miami put the pieces together and started raising a stink. Newspapers were soon denouncing this "low Yankee trick" and calling for Langmuir's head. This once again put him in a tight spot. On the one hand he wanted to prove that he really could influence hurricanes. On the other hand, taking credit for this storm would saddle him with a lot of liability. He would also have to admit that he didn't know what the hell he was doing, since the storm had gotten away from him. Playing God was turning out to be a drag.

With the hurricane work suddenly under scrutiny, Langmuir shifted Project Cirrus to the New Mexico desert, where there were fewer things to destroy and where he could concentrate on simple rainmaking. Now, this might seem like running away, but Langmuir had no stomach for self-pity and no room in his head for self-doubt. If anything, the results that began trickling in from New Mexico were even more fantastical than his claims of steering hurricanes around.

In one trial run, Langmuir claimed that just two ounces of silver iodide, one lousy dollar's worth, had wrung two hundred billion gallons of rain out of some clouds. When asked how that was possible—how so little starting material could yield so much rain—Langmuir argued that he'd produced a chemical chain reaction in the clouds, and compared his work to another New Mexico miracle, the Manhattan Project. Heck, he said, his chain reactions were probably more powerful. America had a new Storm King, it seemed, and as the months passed, he began taking credit for storms not just in New Mexico but throughout the United States, plus a few in Europe. In his diary, Langmuir called this the most important work of his career—and this from a man who'd won a Nobel Prize. Langmuir eventually resigned from GE to pursue this "experimental meteorology" full-time.

Meanwhile, actual meteorologists were taking a good long look at Langmuir's results—and what they found made them suspicious. For one thing every epic rainstorm he took credit for could plausibly be traced to another cause. The two-hundred-billion-gallon storm in New Mexico, for instance, had coincided with a front sweeping up from the Gulf of Mexico. Even more damning, Langmuir's experiments lacked controls, and he seemed to select clouds that were already ripe and likely to rain anyway. When meteorologists ran their own, independent tests with proper controls and random cloud seeding, they found virtually no extra rain.

The same general criticisms applied to hurricane hunting. Hurricanes changed size and direction on their own all the time, constantly waxing and waning in strength. So it was impossible to tell whether seeding actually *caused* the changes Langmuir took credit for or whether it was all coincidence. In fact, Project Cirrus escaped prosecution for the Savannah debacle largely because a meteorologist with a good memory tracked down reports of a 1906 hurricane that had taken the exact same hairpin turn in the Atlantic.

The debate over the validity of Project Cirrus continued for most of the next decade. Meteorologists continued to pick apart Langmuir's experiments, but no matter what their criticism was, Langmuir always had an excuse, every last time, and his Nobel Prize ensured that he always got the benefit of the doubt with outsiders. He also defended himself in public lectures about weather control, and by all accounts he was a captivating speaker, a dynamic blend of avuncular charm and scientific authority. Indeed, although Project Cirrus ran its course in 1952 and Langmuir passed away in 1957, he managed to stir up enough enthusiasm for weather control that the U.S. government started a new project in 1962 to extend and expand upon Project Cirrus. They called it Project Stormfury.

Stormfury commanded a multimillion-dollar budget and several planes, and to go with this kickass new name, Stormfury scientists developed new approaches for attacking clouds. Rather than fling seeds by hand, pilots now packed silver iodide into aluminum canisters and fired it out of Gatling guns. *Rat-a-tat-tat-tat.* And rather than rely on a single drive-by shooting, ten planes in formation might circle a hurricane for an hour, pumping the storm full of silver.

Piggybacking on this work, the U.S. military decided to invest in a separate weather control project—and, more ominously, to convert this research into new weapons. The idea of "weather warfare" actually grew out of another military project, in Vietnam. By 1966 the military had spent millions of dollars defoliating Indochina to improve visibility for bombing runs. Some generals favored Agent Orange for this, but others thought good old-fashioned fires worked best. Fires in the lush local forests, however, often produced loads of smoke. And as James Espy might have prophesied, rain often followed the fires a day or two later. Of course, this was Vietnam—which is part *rain* forest—so rain wasn't exactly rare there. No one bothered doing any sort of statistical analysis, either, to see whether a correlation existed. But the periodic drenchings did plant an idea in the minds of a few officers. Each year between May and September, monsoon rains sweep through Vietnam, sometimes dumping twenty inches per month. These downpours turned most of the mud roads there into veritable Slip 'N Slides—including the Ho Chi Minh Trail, a vital supply route for the Vietcong that wound through several countries. The American officers figured that if they could make the monsoons even worse, they could paralyze the enemy for several months each year. They called the scheme Project Popeye.

When presented with the plan, President Lyndon Johnson practically whooped in delight, and Popeye kicked off in 1967 with seeding runs over North Vietnam and Laos. Newly elected president Richard Nixon expanded the program in 1969, making it truly bipartisan.

For reasons that remain unclear, the air force did the seeding runs with planes originally designed to haul trash. Overall, the Popeye crews flew 2,602 sorties and discharged 47,409 rounds of silver iodide. And to be fair, the military rain dancers had far more modest goals than those of Irving Langmuir, who'd believed that he could end hurricanes forever and make the entire New Mexico desert bloom. The military just wanted to extend the monsoon slog for a few weeks each year and make the peak rains worse. This would soften the roads into muck, wash out some bridges, maybe even cause a few strategic landslides. Officials also congratulated themselves for conducting a more humane form of warfare, since soaking the enemy with water was surely better than dumping napalm. (Popeye's motto was "Make mud, not war.") What's more, cloud seeding was covert. Unless the Vietcong took samples of rainwater and tested for silver iodide, they'd never know.

But the wall of secrecy surrounding Popeye started crumbling in 1971, when the *Washington Post* obtained a classified memo that mentioned it. The Pentagon Papers, leaked that same year, also alluded to it, and the *New York Times* ran an exposé on weather warfare in July 1972. Two days later the air force suspended all rainmaking in Southeast Asia. Despite repeated inquiries from Congress and the press over the next few years, Pentagon officials refused to say anything more. (As one wag noted, this inverted the usual state of affairs rather neatly: someone was finally doing something about the weather, but no one was talking about it.)

A dogged Rhode Island senator named Claiborne Pell finally hauled several Pentagon officials into a hearing in 1974. Pell agreed that drenching people with rain was more humane than bombing them. But cloud seeding was such a crude tool, he argued, that any floods or landslides would hit civilians just as hard as soldiers. Indeed, Pell grilled the officials about a series of floods that had devastated North Vietnam in 1971. Had cloud seeding made them worse? Pentagon officials denied it, saying that they never generated enough rain to cause a flood. But that admission only brought up other, more awkward questions. How much extra rain had Project Popeye produced, then? A few inches per month, tops, officials said. In fact, they doubted that the Vietcong had even noticed their efforts, since it's kind of hard to tell the difference between, say, twenty inches per month and twenty-two. So why, Pell asked, did the military spend $21.6 million ($130 million today) if the program wasn't really working? We had to try something, sir.

Probably not coincidentally, government support for weather control collapsed after such information came to light. (And not a minute too soon: it later came out that some Pentagon advisors were already pushing to expand weather warfare into the broader and even crazier realm of "environmental warfare." Ideas included cutting holes in the ozone layer over hostile countries and triggering earthquakes from afar.) Technically, the uproar over Popeye shouldn't have affected Project Stormfury, which had no military aims; scientists there simply wanted to relieve droughts and defuse hurricanes, far more noble pursuits. But Stormfury was tarnished by association and fell into disrepute. It didn't help that the burgeoning environmental movement of the 1970s made meddling with the weather seem hubristic, if not immoral: after billions of years of practice, Mother Nature probably knew what she was doing better than we did.

Politics aside, support for weather control also dwindled because it just didn't work very well. Sure, scientists believe they can do modest things like clear fog at airports or goose certain types of clouds to squeeze out a little more rain. And scientifically, the work wasn't a waste: meteorologists learned an awful lot from the data they gathered. Unfortunately, much of what they learned undermined the rationale of their experiments. With hurricanes, for instance, scientists had counted on the silver iodide interacting with supercooled water to release heat. But their data revealed that hurricanes actually contain very little supercooled water. This meant that (unlike with clouds) the silver iodide had no raw material to start the chemical chain reactions. The dramatic behavior they sometimes observed after seeding hurricanes, then, was probably just a coincidence.

In the decades since Popeye and Stormfury, dreams of weather control have never quite faded. In 1986, the Soviet Union seeded the hell out of any clouds that passed over Chernobyl, in order to drain them before they reached Moscow and pelted the populace there with radioactive drizzle. Similarly, the Chinese weather service reportedly strafed every last cloud around Beijing in the summer of 2008 to ensure clear skies for the Olympics. Most meteorologists, however, continue to take a dim view of such enterprises. They're especially adamant that we can't do much to stop hurricanes and other violent storms. That's a hard, existential truth to accept, because it's tantamount to admitting how vulnerable we are, how impotent and helpless. In the end, though, you can't cheat nature: the atmosphere is far more powerful than we can readily comprehend.

During his lifetime Irving Langmuir spoke with great eloquence about our duty to take control of the weather, and his charisma won thousands to his cause. But for me, the most profound truth to emerge from several decades of weather control research came from a lowly Stormfury pilot. "It was disappointing to concede we couldn't really do it," he said years later. "But the storms were so big, and we were so small."

———◦———

As hopes for weather control faded, meteorologists consoled themselves that at least their ability to predict the weather was getting pretty good. In fact, after the first computers and weather satellites appeared in the mid-1900s, they had more reason than ever for optimism. Every day in every way, weather forecasting was getting better and better. What meteorologists didn't know—couldn't know—was that the very computers that filled them with such hope would soon double-cross them, by proving precise weather forecasts to be impossible.

An artist's impression of Richardson's Forecast Factory (© François Schuiten).

Lewis Fry Richardson's football-stadium-sized weather forecasting center used human "computers" to predict the weather.

Meteorologists first embraced computers in the early 1900s, back when the word "computer" meant "some sap with a slide rule who crunched numbers all day"—literally, someone who computes. The idea of using computers in weather forecasting originated with English mathematician Lewis Fry Richardson, who sketched out a grandiose scheme for the ultimate weather forecasting center. It would consist of a spherical dome several dozen stories tall, with computers seated in tiers around the inside. The inner surface of the dome would be painted with a map of the globe, with the Arctic up top and the Antarctic below, and all day every day Richardson's computers would stare at lists of numbers and calculate data. The calculations would draw upon seven equations that Richardson used to model the atmosphere, and each worker would focus on one narrow aspect of the weather in one specific part of the world—tracking, say, humidity fluctuations in Inner Mongolia over and over and over. Finally, each computer would send these figures via pneumatic tubes to a master controller, who would stand on a panopticon pillar in the center and assemble them into an omniscient forecast for the globe.

Richardson estimated he would need 64,000 computers to provide real-time forecasts; imagine an NFL stadium full of people totting up figures and muttering. But to prove the idea worked, at least in theory, Richardson ran a pilot calculation in 1916. He did this work under the most inauspicious circumstances possible, as an ambulance driver in France during World War I. The forty-five members of his unit spent most mornings, evenings, and afternoons peeling shattered soldiers out of the mud and hauling them off to hospitals where, like as not, they'd die anyway; collectively, they transported seventy-five thousand patients during the war. Richardson did weather calculations as a way to keep sane. He'd spread his papers out on a desk of hay bales and work until his mind went numb. Other drivers called him the Professor.

Richardson's pilot calculation focused on a day six years earlier, May 20, 1910. This had been a so-called Balloon Day in Europe, when hundreds of meteorologists had sent up balloons and kites to gather weather data across the continent. Their efforts gave Richardson accurate, hour-by-hour information on the state of the weather that entire day, allowing him to check his theoretical calculations against reality. (May 20 also coincided with the return of Halley's Comet—in all, a glorious, carefree day of science and international camaraderie that must have seemed several lifetimes removed from the trenches of France.) Even restricting himself to one day proved overwhelming, however, so Richardson further narrowed his focus to one variable in his seven equations—the air pressure between 4 a.m. and 10 a.m. over a small parcel of Bavaria. The work still took weeks if not months, and he almost had to start over when he lost his notebook during the Battle of Champagne. (Someone later found it beneath a pile of coal.) And what did all that toil earn the Professor? Basically nothing. Beyond being six years late, his "forecast" wasn't even close. According to his numbers, the barometers that morning in Bavaria should have registered a huge drop in pressure. In reality, they'd barely fluttered. To his credit, Richardson published a complete account of his failure in 1922.

However futile Richardson's efforts might have seemed, the next generation of meteorologists quickly forgot the moral of his story and raced ahead with forecasting schemes of their own. They were especially excited when digital computers — actual electronic gizmos — appeared on the scene in the 1940s. In fact, no less an authority than John von Neumann, architect of the first electronic computers, declared ENIAC and other such devices ideal for weather forecasting. This is our year!

One early adopter was Edward Lorenz. Lorenz had earned a degree in mathematics in 1938, but he caught the weather bug after a stint forecasting for the army during World War II. He stuck with it afterward and by the early 1960s had landed a job in the meteorology department at MIT. There, he studied physical models of the atmosphere, including so-called dishpan experiments. These involved swishing water around in a shallow dish with a pencil; after a while, the fluid would erupt into swirls and eddies, a phenomenon known as turbulence.* Surprisingly, the dishpan experiments did provide some insights into the turbulent motion of weather fronts. Still, the setup was pretty crude, and Lorenz hoped that computers would provide more sophisticated results.

Everyone knows how gargantuan and slow ye olde computers were, and kitchen appliances nowadays have more memory than the mainframes that landed Neil Armstrong on the moon. What you don't often hear is how loud early computers were, or how unreliable. Lorenz worked on something called a Royal McBee LGP-30, which whirred like an engine and broke down at least once a week. Its bulky metal frame took up most of his office — it was like shoving an armoire in there — and its vacuum tubes pumped out absurd amounts of heat. The local forecast was always hot and sticky when the McBee was running.

Lorenz wanted to study global weather patterns on the McBee, similar to what Lewis Fry Richardson had proposed, albeit with twelve equations instead of seven. But while Richardson had hewed to real weather data, Lorenz let his imagination roam. He basically invented his own little world where he could alter the temperature, pressure, and wind speed at will, like a demigod. He'd then set everything in motion and observe how the weather evolved day by day over virtual months or years. To you or me, the results of these simulations would have been inscrutable. We're used to weather maps on television with patches of red or blue and ripples of clouds. Lorenz got nothing but printed tables of figures. But he could translate those numbers mentally into rain or sunshine when he read them over — like a musician reading squiggles and dots on paper and hearing a symphony.

One winter morning in 1961, Lorenz revisited a simulation he'd run a little while before. It had shown some interesting behavior, and he wanted to extend the run for longer this time, to see what else emerged. Rather than rerun the entire sequence, though, and risk a core meltdown with the McBee, he took a shortcut and started in the middle. He pick-sixed a line of numbers from the printout he had — the row starting with 0.506 looked good — typed the figures in, and left to grab some coffee while the McBee buzzed.

Lorenz returned an hour later and sat down to study the meteorological score. It immediately looked funny. Since he was repeating part of the previous run, he expected to see identical numbers line by line at the start. Instead, when he compared the first and second runs, he saw the numbers diverging. It happened slowly at first, but grew more and more pronounced with each step. Pretty soon, he couldn't ignore it anymore. It was like one of those schizoid days in March when it's seventy-five and sunny at noon and sleeting at two o'clock — that's how different the two runs looked.

Lorenz grumbled. Same input, same equations, different results.

That could mean only one thing: a short. And like a grouchy dad on Christmas, who has to check an entire string of tree lights for a single dud bulb, Lorenz and a technician would now have to spend the next several hours going over every last relay and switch in the McBee.

Before he ruined his afternoon, though, Lorenz decided to think for a moment. He soon realized something. To save memory, old computers often truncated numbers. (Remember Y2K?) The McBee worked with six decimal places in its internal memory; but when it printed those figures out, it rounded to three. So while Lorenz had inputted 0.506 to start the new run, the true value at that point had been 0.506127. Could that explain the divergence?

Lorenz doubted it. This difference amounted to one part in five thousand, just 0.02 percent. How could such a tiny fraction, a rounding error, have blown up the whole simulation? But as he went over the numbers again, Lorenz started to doubt his doubt. It seemed that this tiny swerve at the start really had overwhelmed the final results.

According to traditional thinking, this made no sense. In any well-behaved system, virtually identical inputs should lead to virtually identical outputs. I mean, imagine dropping an apple and watching it fall to the dirt. If you pick it up, pull out a pocketknife, and shave off 0.02 percent of its weight, you don't expect the apple to suddenly start hovering the next time you drop it. It's pretty much the same apple, so it should behave the same way. But in Lorenz's universe, the 0.000127 he'd shaved off had changed everything. Somehow, when it got run through the wringer of his twelve equations, that one part in five thousand had swelled in magnitude and flipped the weather completely. The interaction of all those equations made things chaotic.

Although intrigued, Lorenz reminded himself that this was still just a simulation; it might not have anything to do with real weather. Yet the more he thought about this glitch, the more profound it seemed. Meteorologists back then were always boasting about how next year's McBee or next year's weather satellite would make precise forecasting possible. But maybe this was moonshine. Maybe churning through more data wouldn't help. Maybe there were simply too many variables to account for. Maybe turbulence and unpredictability were intrinsic features of the atmosphere.

Lorenz downplayed these suspicions at first, merely cracking jokes to colleagues that at least they had an excuse now for bungling the weather report on the six o'clock news. Deep down, though, he felt a frisson of excitement, and he decided to pursue this line of thinking. It led to some strange places: Hell, after a few months he wasn't even sure whether he was doing math or science anymore. He wasn't proving theorems like a mathematician would; he was just tracking trends in tables of data. But this wasn't real data, either, from real science experiments, just simulations on the McBee. And because it was such odd work, Lorenz had difficulty getting it published. He had to resort to subterfuge sometimes, like tacking empty paragraphs about weather forecasting onto his papers, when what he really cared about was the underlying math. Even then he often settled for publishing in obscure places, like Swedish meteorology journals. Not exactly where MIT professors expect to publish.

Gradually, though, as more and more forecast models foundered in the 1960s, Edward Lorenz found his audience. He sat on the advisory board for Project Stormfury, and his increasing skepticism about weather prediction—much less weather control—almost certainly helped kill off that project. In the 1970s something finally tipped, and Lorenz's ideas about the chaotic nature of the weather became mainstream. Lorenz also did himself a tremendous favor by distilling his insights into one of the century's most captivating metaphors. It first appeared in the title of a paper he wrote in 1972, "Predictability: Does the Flap of a Butterfly's Wings in Brazil Set Off a Tornado in Texas?" We now refer to the tendency of small, inconsequential differences to explode in complexity and become very consequential indeed as the butterfly effect.

Other scientists and mathematicians expanded on Lorenz's work in the 1980s, and today we recognize him as a pioneer in what's called chaos theory, a field that reaches far beyond the weather. You can use chaos theory to trace the shapes of mountains and river deltas; to explain why sewage flowing through pipes suddenly turns turbulent; to parse the sawtooth ups and downs of commodities prices; even to predict when the genetically engineered dinosaurs in your theme park will go on a rampage. These seem like totally disparate things on the surface, but they all share underlying similarities, including the tendency to flip from well behaved to badly behaved in a heartbeat (a heartbeat being another sometimes-chaotic phenomenon). Because of its incredible reach, some historians have nominated chaos theory as one of the three true scientific breakthroughs of the past century, alongside relativity and quantum mechanics. If that judgment holds up, our descendants might someday speak of a patent clerk named Einstein and a weatherman named Lorenz in the same breath.

Regardless of its place in history, chaos theory did expose the futility of trying to predict the weather in any remotely precise way. Now, that's not to pick on meteorologists. Fancy new satellites and supercomputers can (usually) nail down a weather forecast a few days ahead of time—a crucial window not only for planning picnics but for warning us about deadly storms. (It's thanks to meteorologists that the odds of dying in a hurricane nowadays are 1 percent of what they were in 1900.) But weeklong weather predictions get sketchy, and *Poor Richard's*-style forecasts that look months into the future are little better than hoodoo. We can predict eclipses decades in advance, but when it comes to the weather there are simply too many pockets of air colliding with too many bumps on Earth's surface to track them all—too many butterflies flapping their wings and kicking up tornadoes.

For the first time in this book, then, our trusty old gas laws come up short. They provided a good, solid framework for everything from flying hot-air balloons to chilling an icebox. And they do a decent job of explaining many basic features of the weather, like the role of water vapor. Ultimately, though, the rush of gases across a spinning planet gets so frenzied that nice, clean, simple volume-temperature-pressure relationships can't keep up—they're left gasping. Before we know it, what should have been a nice summer breeze snarls with turbulence, and those fluffy white clouds above take on a sinister cast. Chaos always wins.

Lorenz certainly wasn't the first person to recognize how complicated the weather is; Lewis Fry Richardson could have told him that. But Lorenz made us confront the fact that we might never be able to lift our veil of ignorance—that no matter how hard we stare into the eye of a hurricane, we might never understand its soul. Over the long run, that may be even harder to accept than our inability to bust up storms. Three centuries ago we christened ourselves *Homo sapiens,* the wise ape. We exult in our ability to think, to know, and the weather seems well within our grasp—it's just pockets of hot and cold gases, after all. But we'd do well to remember our etymology: *gas* springs from *chaos,* and in ancient mythology chaos was something that not even the immortals could tame.

Interlude: Rumbles from Roswell

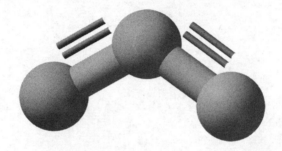

Ozone (O_3)—currently 0.1 parts per million near the ground, where you inhale a quadrillion molecules every time you breathe; more than one part per million in the stratosphere (where you shouldn't be breathing the air!)

In addition to using computers, meteorologists last century took advantage of new balloon technology to explore the workings of the atmosphere, especially its upper reaches. And just as with computers, the balloon projects led to several important insights into how air works—as well as, in one memorable case, to no small amount of embarrassment for the scientists on the ground.

It all started one morning in June 1947 when a ranch foreman named Mac Brazel came across a trail of metal and plastic debris after a thunderstorm. Now, Brazel had no intention of igniting a half century of hysteria and conspiracy theories; he just wanted to clean up the damn ranch. So rather than leave the scraps there and risk his sheep chewing on them, he gathered them up, tossed them into a shed, and tried to forget about them.

Some of the scraps that Mac Brazel (not pictured) discovered on the ranch. (Photo courtesy *Fort Worth Star-Telegram* collection, special collections, the University of Texas at Arlington libraries)

Except, the more he thought about it, the more the scraps bothered him. He worked on a ranch near a few military bases in New Mexico, and scientists around there were always launching missiles and weather balloons that came crashing back down on people's land. He'd in fact found downed weather balloons twice before. But this time was different. The crash landing had gouged out deep grooves in the earth, which seemed impossible for a soft balloon. And the shards of plastic and metal didn't seem like balloon material. Most unsettling of all, the wreckage included a few short wooden beams with purple squiggles on them, like writing—but writing in no earthly language he knew of.

A few days later Brazel showed the scraps to his neighbors. They in turn told him some rumors they'd heard lately about unidentified flying objects near their lands. This spooked Brazel good, so he visited the local sheriff in Roswell, seventy-five miles away, on July 7. The sheriff in turn called up officials at a nearby army air force base.

When they arrived at the shed, the air force officials examined the scraps and tried to reconstruct the thing; they soon gave up, baffled. They also tried to cut the metal bits with Brazel's knife and burn the scraps with matches, and failed on both accounts. They finally examined the purple squiggles, which they began referring to as "hieroglyphics." At this point they decided to confiscate the whole mess.

Thanks to Brazel the local gossip mill had been churning for days by this point. But rather than keep mum, the air force issued a spectacularly boneheaded press release claiming that "rumors regarding the 'flying saucer' became a reality yesterday." A newspaper story made similar claims. To be sure, phrases like "flying saucer" and "unidentified flying object" had more neutral meanings back then, but it didn't take long for people's imaginations to endow them with very specific meanings indeed.

All the scuttlebutt still probably would have died down, except that senior air force officials swooped in and demanded the retraction of the press release; one actually drove around to local newspapers and radio stations and snatched up paper copies. This got even skeptics thinking hard about conspiracy. What was the air force scared of? What were they hiding? People grew more suspicious still when the air force insisted that all the scraps had come from a weather balloon—obvious bull sugar. And in fact, we can now say for certain that the air force lied about this: that was no weather balloon Mac Brazel found. Unfortunately, what the military *was* lying about probably isn't what you're hoping for—unless you're a spy buff with some pretty esoteric knowledge about the atmosphere.

The whole Roswell fiasco started with an earthling named Maurice Ewing, a geophysicist at Columbia University who did contract work for the military. Like every other red-blooded American then, Ewing dreaded the prospect of the Soviet Union acquiring the Bomb. But in those days before satellites and fallout detectors, we had no idea what the Soviets were up to. So he started thinking about other ways to spy on the Reds. He finally hit upon a way to eavesdrop on atomic blasts from afar by suspending microphones in a region of our atmosphere called the sound channel, which is located roughly nine miles up in the sky.

To understand Ewing's idea, you have to know three things about sound. First, sound moves faster in warm air than in cold air. That's because sound depends on molecules knocking into one another. It's pretty slapstick, actually. When someone speaks, the air molecules leaving her mouth crash into nearby molecules of air. These careen into a second layer of molecules, which blunder into a third, and so on, until the noise stumbles into your ear.* The key point here is that air molecules at high temperatures are moving faster than air molecules at low temperatures. And since sound is essentially a relay race of air molecules, the faster-moving molecules in warm air can transmit sounds more quickly: in air at 0°F sound travels at 718 miles per hour; at 72°F it jumps to 772.

The second thing to know is that sounds don't always follow

straight lines; they bend in certain circumstances. Specifically, if there are layers of warm and cold air around, sound waves always bend toward the slower layer—toward the colder air. This bending is known as refraction.

To see refraction in action, imagine a trumpet player standing in the end zone of a domed football stadium. Imagine too that the stadium's air conditioners are struggling to keep the place cool: there's a nice layer of cold air near the ceiling but the field below is bathed in warm air. Because of refractive bending, any tootles on the trumpet will curve upward toward the cooler air. This means that someone standing in the opposite end zone will have trouble hearing anything, since the sound will sail over her head. Conversely, imagine a game later in the season. Now the stadium's heaters are struggling, leaving the dome with a layer of warm air up top and cold air below. In this case the trumpet notes might start to rise—but will soon get bent back toward the ground, making them easy to hear. Again, sound always bends toward colder air.

The third thing about sound involves the temperature profile of our atmosphere. We all know that air gets colder as you rise, which explains why mountaintops near the equator can be snowcapped. By around 45,000 feet, the air temp drops to $-60°F$, which slows the speed of sound to 672 miles an hour. And just as you'd expect, noises outdoors tend to curve upward toward this cooler air. This explains why early balloonists could hear dogs barking and roosters crowing with such clarity. The atmosphere was actually funneling noise up toward them.

But the air cools as you rise only up to a point—that point being around 60,000 feet, when ozone starts to appear. Ozone absorbs ultraviolet light that would otherwise scramble our DNA; life never could have oozed out of the ocean and onto land without it. And in absorbing ultraviolet light, ozone warms up. All the ozone in the atmosphere, if collected and pressed together, would form a shell just an eighth of an inch thick. But it absorbs ultraviolet light so well that even this trifling amount of gas can warm the air at 150,000 feet to a balmy 32°F. Overall, then, our air forms a sort of temperature sandwich: there are two layers of warm air (one near the ground, one at around 150,000 feet), with a slice of cold air in the middle.

The notorious "Flying Saucer" headline in the *Roswell Daily Record*, July 6, 1946.

And here's the payoff: this temperature profile sends sounds on a wild ride. Consider a hunter on the ground who blasts a shotgun. Per the discussion above, that sound will rise, curving toward the cooler upper air. But the thing is, sounds don't just stop when they reach this layer. They have momentum, they keep going. So after passing through the cold layer at 45,000 feet, the sound will inevitably run into the ozone-warmed air above it. And because sound always bends away from warm air and toward cool air, the shotgun noise will actually do a gentle U-turn at this point and begin falling like an arrow. In other words, ozone reverses the sound's direction, as if it bounced off a wall.

What happens next is even stranger. After it starts falling, the sound still has a fair amount of momentum. So it plows right through that cold layer at 45,000 feet and heads toward the ground. But what happens as it approaches the ground? It encounters a layer of warm air. And because sound always (say it with me) bends away from warm air and toward cold air, the majority of the sound energy will turn tail again and start to rise. But this of course sends it on a recollision course with that upper layer of ozone-warmed air. Whereupon it pulls a third about face, and starts to sink. And it keeps sinking—until it meets the warm air near the ground and bounces back toward the sky again. In other words, the sound gets stuck in a loop. It keeps rising and falling, rising and falling, oscillating around that cold layer of air. That's why this cold layer is called the sound channel, because sounds get swept toward it and have trouble escaping.

There are a few caveats worth noting about the sound channel. First, only pretty intense noises have enough energy to rise that high and get sucked into it. It's not like your sweet whispered nothings from last night are still bouncing around the stratosphere, thank God. What's more, the most intense noises,* after their initial U-turn in the sky, do sometimes have enough energy and momentum to push through the layer of warm air near the ground and strike the ears of listeners below. We've encountered this already with Mount Saint Helens. Remember that people near the eruption heard nothing, while people far away got clobbered with noise. That's because the boom initially curled upward toward the cooler air, sailing over the heads of those nearby and creating a sixty-mile-wide "sound shadow." But the boom nose-dived when it hit those warmer pockets of air above, allowing people farther away to hear. Something similar happened with the nuclear bomb at Hiroshima. Survivors near the epicenter spoke of the *pika,* the flash, while those farther away recalled the *pika-don,* the flash-boom.

Maurice Ewing first worked out the physics of the sound channel in 1944.* It seemed little more than a novelty, though, until he realized something else. He now understood what happened to noises that originated above or below the channel—they got funneled into it. But what about noises that originated *within* the channel? How would they behave?

Consider a shotgun blast again, but this time at 45,000 feet, at the temperature nadir. Like all sounds, no matter where they originate, the noise from this blast will initially start to spread in all directions. And in spreading out like this, sounds usually dissipate, weaken. But something unusual happens around this specific height. No matter what direction the sound waves go, up or down, they encounter warmer air and get nudged back toward the center. As a result, sounds that start within the sound channel don't spread out much—meaning they don't weaken. They're therefore audible at much farther distances than normal. They're effectively magnified.

In 1947 Ewing realized that this effective magnification of sounds offered a clever way to spy on the Soviets. Now, the Soviets weren't going to explode nuclear weapons nine miles up in the sky—that's awfully high. But Ewing knew that mushroom clouds often do rise that high. Mushroom clouds are pockets of hot gas that knock other air molecules around. Knocking air molecules around is basically the definition of sound, and Ewing hoped that Soviet mushroom clouds would raise enough of a ruckus at nine miles high for him to hear it halfway around the world. All the air force had to do was send balloons with microphones into the sound channel to eavesdrop. The air force called the scheme Project Mogul.

Ewing was pretty optimistic about Project Mogul at first, but when he began running tests at Alamogordo Army Air Field in New Mexico in early 1947, he ran into several problems. One involved keeping balloons at a constant altitude, since sunlight warmed the balloon envelope. This in turn warmed the gas inside and caused the balloon to rise out of the sound channel. Ewing's team countered this tendency by using transparent balloons, which allowed sunlight to stream through. (Ewing ordered them from the same company that made the first balloon figures for the Macy's Thanksgiving Day Parade. When his assistants saw the transparent balloons, they immediately thought of something else: titanic condoms.)

Another problem involved tracking the balloons, since they wandered aimlessly with the wind. Ewing proposed tracking them with radar, but the equipment at Alamogordo had trouble finding these tiny targets at high altitudes. So the scientists decided to send up not one but thirty balloons at once; they were yoked together in a column sixty-five stories tall, more than twice the height of the Statue of Liberty. They also added radar reflectors to the balloon column, metal surfaces that helped redirect the radar waves back toward the ground. Each reflector looked something like a metallic box kite, and Project Mogul in fact contracted with a toy company to make them. Because the scientists didn't care about aesthetics, the toy company bound the reflectors together with Elmer's glue and tape. And because tape was scarce due to lingering wartime shortages, the company dipped into a stock of wacky novelty tape it had on hand—tape covered in purple, squiggly hieroglyphics.

Reenactment of the Project Mogul balloon experiment. (Photo courtesy Joe Nickell/*Skeptical Inquirer*)

As you probably guessed, these ungainly columns of metal, plastic, and rubber accounted for many of the "unidentified flying objects" that invaded the skies of Roswell in 1947. When aloft, the columns moved in mysterious ways, with different parts snaking back and forth at different times, depending on the wind. The radar reflectors glinted eerily in the moonlight as well, and when the columns crashed down, the metal gouged the earth and produced far more debris than any weather balloon could have.

This tendency to strew debris about became a headache for Maurice Ewing. Crazy as it sounds, Project Mogul received the same ultra-double-secret classification that the Manhattan Project had. Not even the folks at Roswell Army Air Field ninety miles away knew about it, which meant that Ewing's team had to scramble to retrieve every scrap from every one of the 110 flights they launched. Most of the time they found the downed balloons easily enough; when they lost one, they actually listened to radio reports of UFO sightings for leads. But some balloon columns escaped—including the one that crash-landed on Mac Brazel's ranch.

Given all the hoopla that followed, in later years Brazel said he regretted not keeping his shed locked and his trap shut. But for whatever reason, people across the globe seized upon his story, and the debris he'd found in the dirt somehow acquired otherworldly powers. The military's response only fueled people's suspicions, and Roswell soon metastasized into the phenomenon we know today.

Meanwhile, Project Mogul continued in secret for another few years, and a few accounts claim that Mogul balloons did detect Joe-I, the first Soviet nuclear weapon test, in August 1949. But so did other, cheaper, more reliable methods, such as sending airplanes aloft to scour the skies for radioactive dust. After years of marginal results, the air force finally shuttered Mogul in 1950.

At this point, with Mogul in the dustbin of history, the military could have come clean. Paranoid to the end, however, officials continued to stonewall and insist on the silly weather balloon story. Apparently the threat of the Soviet Union loomed so large in their imaginations that they preferred to let rumors about an alien invasion fester rather than tip off the Soviets to even a failed attempt to spy on them. By the time the air force owned up to Project Mogul, in the 1990s, it was too late: the Roswell rumors had taken on a truth of their own.

In a topsy-turvy way, though, history has proved the conspiracy-mongers right. The air force was indeed lying all those years, and it was indeed desperately scanning the skies above Roswell in 1947 — but for portentous rumbles of gas, not alien star cruisers. And to think the whole twisted tale started with an acoustic quirk of our atmosphere, which in turn depended on the energy-absorbing prowess of ozone. By protecting the DNA of landlubbing creatures, ozone arguably did as much as any other gas to accelerate the evolution of life on Earth. And by enabling Project Mogul, ozone also convinced more people than ever of the subject of our next chapter, the existence of life on other planets.

CHAPTER NINE

Putting on Alien Airs

All book long we've seen examples of how vast our ocean of air is and how profoundly it has shaped—and continues to shape—human life. Now it's time to expand our horizons yet again and explore the atmospheres of other planets. Because however rich and rewarding it is to study, Earth's atmosphere is just one example. So what other kinds of air exist in the infinitude of space? What airs do alien life-forms breathe? And what would happen if human beings tried to breathe those airs, too?

Of course, in talking about those other atmospheres, we'll discover new things about our own air as well—including how precious and even fragile it is. The last chapter ended by pointing out that human beings can manipulate our weather only to a limited degree. But I didn't mean to imply that our atmosphere is so overwhelmingly big and that we humans are so underwhelmingly puny that we can't affect our air in any way whatsoever. To the contrary. We might never engineer the weather to suit our whims, but that docsn't mean we aren't changing the climate in other, more significant ways.

————<o>————

It's a shame that the tinfoil-hat/anal-probing crowd has hijacked the debate ever since Roswell, because the study of alien life actually has an illustrious pedigree. Johannes Kepler swore he'd found proof of advanced civilizations on the moon; ditto William Herschel and the sun. Immanuel Kant and Christiaan Huygens wrote at length about extraterrestrials, as did Carl Gauss and Benjamin Franklin. A few thinkers even suggested ways to signal like-minded aliens: grow gargantuan right triangles of wheat in Siberia, or fill huge canals in the Sahara with oil and light them on fire.

Probably the best-known proponent of alien life was Percival Lowell, a wealthy American ambassador and author who grew obsessed with astronomy later in life. Lowell was inspired by the work of an Italian astronomer who claimed in the 1870s to have found a grid of black *canali* crisscrossing the surface of Mars, some of them three thousand miles long. Reading about these *canali,* Lowell thought immediately of the Suez Canal, then the most marvelous engineering project in history. He convinced himself that Martian canals were even grander, and he built his own observatory in Arizona to study them, spending $20,000 ($500,000 today) on the telescope alone. Lowell might have done better to spend $20 on an Italian tutor, who would have told him that *canali* meant not "canal" but the more neutral term "channel." Regardless, Lowell began to study the Martian "waterworks" in earnest; he also claimed to have found evidence for the waxing and waning of "vegetation" each Martian spring and fall. Most astronomers supported Lowell at first, until he began publishing books full of wild speculation about Martian technologies. The last straw was when he claimed to see signs of civilization on Venus, an impossibility given the constant cloud cover there. Better telescopes finally revealed the "canals" to be optical illusions—like a string of black dots that blur into a line from a distance. Biologist Alfred Russel Wallace also pointed out that, from an engineering standpoint, Lowell's claims made no sense. Canals that carried water for thousands of miles would lose every drop to evaporation. Moreover, Wallace noted that these canals never turned or weaved, never deviated around natural features in the landscape, the way you'd expect. If these really were canals, Wallace concluded, they were "the work of...madmen rather than of intelligent beings."

Lesson not learned, astronomers continued to let their imaginations roam over the next century. In the 1970s NASA launched two *Viking* probes to search for life on Mars, and before takeoff, Carl Sagan gathered a gaggle of reporters around a model *Viking* lander and began parading snakes, chameleons, and tortoises past its cameras, to show everyone what he expected to find on the Red Planet. "There's no reason to exclude from Mars organisms ranging in size from ants to polar bears," he declared. No rust-colored polar bears ever turned up.

Here's what we know about the current prospects for life in our solar system. We can rule out some places immediately. With all due respect to William Herschel, his theory about creatures living on the sun was daft. The sun is far hotter even than fire, hot enough that atoms there disintegrate into plasma. Good luck building complicated biomolecules like DNA in those circumstances, much less entire organisms. As for the other object that dominates our sky, the moon seems too cold and especially too dry for life. Perhaps it's bio-prejudice on our part, but a liquid like water seems essential to anything we'd recognize as living, both to provide a medium for chemical reactions and to gobble up free radicals that would otherwise shred organic molecules. Nor does the moon have any air to breathe.

Mars and Venus once showed great promise as abodes for life, but each stumbled for different reasons. The canal contretemps notwithstanding, scientists now believe that Mars did have flowing water at one point, just not anytime recently. It also lost most of its atmosphere a few billion years ago. As with Earth, this atmosphere was the result of volcanic eruptions, and also like with Earth, those early atmospheres probably got blasted into space after asteroid strikes. But unlike Earth, Mars was too small to retain much interior heat. As a result the interior cooled and solidified, and the volcanoes dried up. Mars therefore lost the ability to replenish its atmosphere after the slate got wiped clean. The solidification of its core also killed Mars's magnetic field. That's a big deal because a magnetic field essentially acts like a force field around a planet and deflects the solar wind, a stream of particles from the sun that tends to strip away precious gases. All Mars has for an atmosphere nowadays is a whiff of carbon dioxide, with an air pressure two hundred times less than that on Earth.

(Incidentally, it's a myth that your head would explode if you took off your space helmet in a low-pressure environment like on Mars, or in outer space generally. Your skull is strong enough to withstand that. That said, you wouldn't last long. Given the minuscule pressure, all the water in your mouth and eyes would boil away in seconds. Your body would also shut down as the intense cold turned your brain into a block of ice. Welcome to space, where you can boil and freeze simultaneously.)

Venus, meanwhile, started off as a second Earth, virtually identical except for an orbit closer to the sun. That one difference did our twin in, though. In a sense, Venus has the opposite problem of Mars—too much air. Its ancient volcanoes probably released water vapor and carbon dioxide at about the same rate as on early Earth. But because Venus sat closer to the sun, the temperature there never dropped far enough for the steam to condense into lakes and oceans. It stayed gaseous. And without standing water, the CO_2 never had a chance to dissolve away and form solid minerals; it too remained airborne. Overall, there are roughly the same number of carbon atoms on Venus and Earth, but on Venus there's 200,000 times more carbon-based *gas*. This gives Venus an air pressure comparable to the pressure a half mile deep in the ocean. Worse, carbon dioxide traps heat (it's a greenhouse gas), and the surface of Venus now roasts away at an incredible 860°F, hot enough to melt lead. Don't damn people to hell,* damn them to Venus.

Most astronomers today agree that if life exists anywhere else in the solar system, it's on the moons of Jupiter or Saturn. Because of their distance from the sun, none of these moons get much light or heat, but they do feel strong tidal forces as they orbit their mother planets. Tidal action basically converts gravitational energy into friction, and that friction probably provides enough heat to produce volcanoes and liquid water, at least beneath the moons' surfaces. NASA considers the Jovian moon Europa such a promising candidate for life that, after the *Galileo* probe finished circling Jupiter in 2003, scientists deliberately crashed it into Jupiter rather than risk it falling onto Europa someday and perhaps contaminating it with microbes that hitchhiked from Earth.

As for whether life exists beyond our solar system, scientists have been flip-flopping on the idea for decades. Some think it impossible, while others are convinced it's out there. (As Arthur C. Clarke once said, "Two possibilities exist: either we are alone in the universe or we are not. Both are equally terrifying.") For several reasons, scientific opinion in the past few decades has swung decidedly toward the latter possibility, that the universe must be teeming with life.

For one thing, we now have hard evidence that planets exist around other stars. This work really didn't get going until the 1990s, but astronomers have already located 3,200 of these so-called exoplanets. Usually they detect them by looking for periodic shifts in the light that stars produce. (Especially Doppler shifts, subtle changes in the color of light as the star gets tugged this way and that by the circling planet.) If the star and planet line up the right way, scientists can also look for periodic changes in brightness as the exoplanet slides in front of the star and blocks a wee bit of light—a partial eclipse. It's hard to fathom the precision involved with this work: it's like standing in Maine and searching for a flea on a lightbulb in San Diego. Yet scientists have developed the superpowers to do this. There's no sign of life yet—which would be like spotting individual cells, or even molecules, on that flea. But in most cases scientists can determine the planets' size, mass, and orbiting distance, important first steps.

Alien life also seems more plausible nowadays because we know that several potential building blocks for life—water, methane, ammonia, carbonaceous gases—are all common in space. Astronomers have detected DNA bases and simple amino acids out there as well. Equally important, we now know that life on Earth can thrive in pretty harsh places—underwater volcanic vents, the Dead Sea, a half mile beneath Antarctic ice. The bacterium *Deinococcus radiodurans* can even survive in nuclear waste dumps, at radiation levels three thousand times beyond what would fell a human being. (How? By repairing its DNA very, very quickly. *D. radiodurans* didn't evolve to live in nuclear waste, of course, since nuclear waste dumps don't exist in nature. It evolved to live in extremely dry places, and the DNA damage caused by radioactivity happens to resemble that of extreme dehydration. So if you want to know which species would survive a nuclear holocaust, look to the desert.) Overall, the work on biomolecules suggests that the raw material for life is abundant, and the work on extreme environments suggests that life can get a foothold pretty much anywhere.

Still, all this talk about life on distant planets remains speculative — we're like medieval scholastics arguing about angels — until we scrounge up some actual evidence. And the best evidence out there, short of landing on exoplanets, will come from studying gases in their atmospheres.

Artist's impression of starlight filtering through a distant exoplanet's atmosphere. (Photo courtesy NASA)

To gather this evidence, astronomers first need to find a suitable target — a rocky planet neither too close to its sun nor too far away. They'd then wait for the planet to pass in front of that sun. Most of the sunlight would still get through during this transit, since stars are vastly larger than planets. A small percentage would also get blocked by the planet's body. (Earth would block 0.008 percent of our sun's light, for instance.) What really interests astronomers, though, is the even smaller percentage of light (maybe 0.00005 percent) that would neither get blocked by the planet's body nor miss the planet completely. Instead, this light will get filtered through the gas corona of the planet's atmosphere.

Light and gases interact in a special way. When gases get excited, they often emit light—we saw this already with sodium street lamps and so-called neon lights. But there's more to this light than meets the eye. Although the light emitted by those lamps looks like one uniform color, it's not. It's actually a blend of several different colors.

You can see these individual colors—they appear as thin lines or bands—if you spread this light out with a prism. Hydrogen gas, for instance, emits one bright red band, one soothing aqua band, and a few muted purples. Helium meanwhile emits a gorgeous yellow line, among others. (This bold yellow stripe, in fact, was what allowed astronomers to discover helium in the sun in 1868,* decades before William Ramsay discovered it on Earth.) Every other element on the periodic table emits its own characteristic bands of color as well. Scientists call each element's unique array its emission spectrum.

There's also something called an absorption spectrum. This is essentially the opposite of an emission spectrum: whereas an emission spectrum involves hot gases emitting specific colors of light, an absorption spectrum involves cool gases *blocking* certain colors. Imagine a rainbow, the kind you get when you spread pure white light out with a prism. Now imagine someone coming along with some ink and a fine paintbrush, and blacking out a line here, a line there. That's what absorption spectrums look like.

When astronomers look at the starlight filtering through a distant planet's corona, it's the absorption spectrum they're interested in. That's because every different gaseous compound, like chlorine or water vapor or ammonia, absorbs different bands of starlight. The missing bands of color therefore act as a "fingerprint" for that gas, and by studying the patterns of missing colors, astronomers can infer what gases exist in the exoplanet's atmosphere. Despite the daunting technical challenges, space-based telescopes have already detected water vapor on planets several light-years distant—roughly twenty trillion miles away. Future telescopes should be able to pick out carbon dioxide, hydrogen sulfide, ammonia, methane, and other common gases. (Surveys on Earth suggest that life-forms here can produce more than 600 different gases overall.)

Of course, when searching for alien life, some gases are more helpful than others. Water and carbon dioxide imply the presence of volcanoes but little else. Argon just means that a lot of potassium-40 has been lying around and decaying. Hydrogen or helium probably indicates a baby planet, one too young for life. As for positive signs of life, astrobiologists once thought that oxygen would be a dead (so to speak) giveaway, since living organisms produce the vast majority of O_2 on Earth. Ozone seemed a strong sign, too, since the creation of O_3 requires O_2 as raw material. We've since realized, however, that different planets with different geologies might produce oxygen through nonbiological means. (Intense ultraviolet light can split water vapor into H_2 and O_2, for instance.) Similarly, many astrobiologists in the past proposed searching for methane, a prominent microbial waste product on Earth. But models now suggest that underwater spumes of magma might react with seawater and produce methane as a by-product. Heck, even Pluto acquires a sparse comb-over of methane (and nitrogen) as it approaches the sun, and the odds of life existing there are nil.

Ultimately, no one gas can act as a neon "Life here!" sign. That said, *combinations* of certain gases would be strong evidence. When they mingle in the atmosphere, methane and oxygen tend to attack each other, and their concentrations dwindle in tandem. On a lifeless planet, then, you might find appreciable amounts of one or the other but not both. In contrast, if you do find plenty of both, something must be constantly replenishing them, and it's hard to think of what that "something" might be except life. Finding a strong methane–oxygen spectrum would therefore be akin to finding a fossil, a fossil made of gases.

We can take this one step further. Because while gases like oxygen can help us detect alien "plants" and "animals" and microbes, eventually we're going to want to hunt down intelligent life. The so-called SETI movement (Search for Extraterrestrial Intelligence) has mostly focused on detecting electromagnetic waves from distant planets—alien ham radio. But that approach has some shortcomings, in that it picks up only those civilizations that beam things into space. In other words, it would have missed everything about humankind before about 1905, and would miss completely any planets whose inhabitants focused on telegraph-like technologies. What's more, if current trends hold and broadcasting continues to dwindle in importance, Earth itself might be largely "radio quiet" in another century or two, rendering us largely invisible from afar. Civilizations on other planets might follow a similar pattern, leaving an awfully narrow window for us to eavesdrop.

A better way to S for ETI, then, might be to hunt for alien pollution. To use Earth as an example, alien astronomers would certainly twitter (or chirp or grunt, or whatever they do when they get excited) over the presence of chlorofluorocarbons (CFCs) in our air, since no natural process can make those gases. Astronomers could also infer some things about the well-being of a far-flung civilization based on its pollution. Some pollutants decay after about ten years, while others require hundreds of thousands. So if we saw a mix of short- and long-lived pollutants in a distant atmosphere, we could conclude that industry was active there. If we saw only the longest-lived ones, the conclusion might be grimmer: that the little green men there had annihilated themselves, perhaps by destroying their planet's environment. (In which case they weren't really little *green* men after all.) Looking for alien pollution might seem far-fetched, but when the James Webb Space Telescope launches in a few years, it will have the capacity to detect, on planets several light-years away, CFC concentrations just ten times greater than those on Earth. The next generation will do better still.

Or maybe it won't be CFCs that tip us off—maybe we'll catch a whiff of another exotic gas. Maybe that gas won't be a pollutant, either, but something we can use here to our own benefit. Spin a periodic table and plop your finger down on a few random elements. Perhaps the gas they form together will revolutionize medicine or transportation or metallurgy in ways we can't yet fathom, just as other gases did in the past. And to think, the first inkling of their existence wouldn't come from some R&D lab on Earth, but in the halo of light from a planet millions of millions of miles away.

———◦———

The hunt for life on other planets raises all sorts of highfalutin spiritual questions* about human beings and our place in the cosmos. (Most pressing, would the discovery of intelligent life elsewhere automatically make us less special?) Unfortunately, rising levels of greenhouse gases here on Earth are also making the habitability of distant planets an uncomfortably practical concern: we might need them as a refuge someday.

Similar to radioactivity, it's important to know that greenhouse gases per se aren't evil. Think of them like cholesterol. Your body actually needs some cholesterol to sheathe your brain cells and manufacture certain vitamins and hormones; it's only when cholesterol levels rise too high that the trouble starts. Same with greenhouse gases.

Greenhouse gases got their name because they trap incoming sunlight, albeit not directly. Most incoming sunlight strikes the ground first and warms it. The ground then releases some of that heat back toward space as infrared light. (Infrared light has a longer wavelength than visible light; for our purposes, it's basically the same as heat.) Now, if the atmosphere consisted of nothing but nitrogen and oxygen, this infrared heat would indeed escape into space, since diatomic molecules like N_2 and O_2 cannot absorb infrared light. Gases like carbon dioxide and methane, on the other hand, which have more than two atoms, can and do absorb infrared heat. And the more of these many-atomed molecules there are, the more heat they absorb. That's why scientists single them out as greenhouse gases: they're the only fraction of the air that can trap heat this way.

Scientists define the greenhouse effect as the difference between a planet's actual temperature and the temperature it would be without these gases. On Mars, the sparse CO_2 coverage raises its temp by less than 10°F. On Venus, greenhouse gases add a whopping 900°F.* Earth sits between these extremes. Without greenhouse gases, our average global temperature would be a chilly 0°F, below the freezing point of water. With greenhouse gases, the average temp remains a balmy 60°F. Astronomers often talk about how Earth orbits at a perfect distance from the sun—a "Goldilocks distance" where water neither freezes nor boils. Contra that cliché, it's actually the combination of distance and greenhouse gases that gives us liquid H_2O. Based on orbiting distance alone, we'd be Hoth.

By far the most important greenhouse gas on Earth, believe it or not, is water vapor, which raises Earth's temperature 40 degrees all by itself. Carbon dioxide and other trace gases contribute the remaining 20. So if water actually does more, why has CO_2 become such a bogeyman? Mostly because carbon dioxide levels are rising so quickly. Scientists can look back at the air in previous centuries by digging up small bubbles trapped beneath sheets of ice in the Arctic. From this work, they know that for most of human history the air contained 280 molecules of carbon dioxide for every million particles overall. Then the Industrial Revolution began, and we started burning ungodly amounts of hydrocarbons, which release CO_2 as a by-product. To give you a sense of the scale here, in an essay he wrote for his grandchildren in 1882, steel magnate Henry

Bessemer boasted that Great Britain alone burned fifty-five Giza pyramids' worth of coal each year. Put another way, he said, this coal could "build a wall round London of 200 miles in length, 100 feet high, and 41 feet 11 inches in thickness—a mass not only equal to the whole cubic contents of the Great Wall of China, but sufficient to add another 346 miles to its length." And remember, this was decades *before* automobiles and modern shipping and the petroleum industry. Carbon dioxide levels reached 312 parts per million in 1950 and have since zoomed past 400.

People who pooh-pooh climate change often point out, correctly, that CO_2 concentrations have been fluctuating for millions of years, long before humans existed, sometimes peaking at levels a dozen times higher than those of today. It's also true that Earth has natural mechanisms for removing excess carbon dioxide—a nifty negative feedback loop whereby ocean water absorbs excess CO_2, converts it to minerals, and stores it underground. But when seen from a broader perspective, these truths deteriorate into half-truths. Concentrations of CO_2 varied in the past, yes—but they've never spiked as quickly as in the past two centuries. And while geological processes can sequester CO_2 underground, that work takes millions of years. Meanwhile, human beings have dumped roughly 2,500 trillion pounds of extra CO_2 into the air in the past fifty years alone. (That's over 1.6 million pounds per second. Think about how little gases weigh, and you can appreciate how staggeringly large these figures are.) Open seas and forests will gobble up roughly half that CO_2, but nature simply can't bail fast enough to keep up.

Things look even grimmer when you factor in other greenhouse gases. Molecule for molecule, methane absorbs twenty-five times more heat than carbon dioxide. One of the main sources of methane on Earth today is domesticated cattle: each cow burps up an average of 570 liters of methane per day and farts 30 liters more; worldwide, that adds up to 175 billion pounds of CH_4 annually—some of which degrades due to natural processes, but much of which doesn't. Other gases do even more damage. Nitrous oxide (laughing gas) sponges up heat three hundred times more effectively than carbon dioxide. Worse still are CFCs, which not only kill ozone but trap heat several thousand times better than carbon dioxide. Collectively CFCs account for one-quarter of human-induced global warming, despite having a concentration of just a few parts per billion in the air.

And CFCs aren't even the worst problem. The worst problem is a positive feedback loop involving water. Positive feedback — like the screech you hear when two microphones get acquainted — involves a self-perpetuating cycle that spirals out of control. In this case, excess heat from greenhouse gases causes ocean water to evaporate at a faster rate than normal. Water, remember, is one of the best (i.e., worst) greenhouse gases around, so this increased water vapor traps more heat. This causes temperatures to inch up a bit more, which causes more evaporation. This traps still more heat, which leads to more evaporation, and so on.* Pretty soon it's Venus outside. The prospect of a runaway feedback loop shows why we should care about things like a small increase in CFC concentrations. A few parts per billion might seem too small to make any difference, but if chaos theory teaches us anything, it's that tiny changes can lead to huge consequences.

Life will go on after climate change, of course — there's just no guarantee that *human* life will survive. So, assuming we don't want to drive ourselves extinct, what can we do? I'm dubious that we'll ever stop polluting voluntarily and revert to simpler lifestyles; we like meat and cell phones and fast transportation a bit too much. Fines or taxes, on the other hand, probably would curb consumption. Some economists also propose setting up a cap-and-trade system where those who adopt cleaner technologies, such as solar power, would earn "carbon credits" to swap for cash. In support of this, they note that a cap-and-trade system helped curb acid rain in the 1980s, largely eliminating that problem. Still, greenhouse gases are more complicated to regulate than acid rain. Curbing acid rain involved monitoring just a few gases (mainly sulfur dioxide and some nitrogen oxides). With greenhouse gases we're talking a few dozen species, including the ubiquitous CO_2, which ratchets up the complexity. And in a larger sense, even if we did curb the release of greenhouse gases, we'd likely have trouble controlling the damage that's already taking place. Because of the positive feedback loop with evaporation and water vapor, global warming has acquired an ominous momentum. We can't just warm the planet a little and hit pause.

For me, climate engineering—taking deliberate steps to cool our atmosphere—seems like the only realistic solution. To be clear, in any sort of objective framework climate engineering also seems desperate and insane. We can't really test the idea on anything less than a planetary scale, and there's a decent chance we'll foul things up royally and make matters worse. But as much as I believe in the law of unintended consequences, I believe even more strongly in the consistency of human nature over time. And given that laziness and shortsightedness have dominated our behavior in the past, I don't see why they won't dominate our behavior in the future as well. I don't mean to sound gloomy about human beings—some of my best friends are people. But we have our flaws, and trying to deal with climate change exposes the worst of them. In contrast, coming up with a technological fix for the problem, while not easy, exploits what humans do well—rally around a cause when things get desperate, then start building shit.

One approach to climate engineering involves capturing gaseous carbon dioxide and transforming it into solid carbon. For instance, when ants burrow into the soil and build nests, they produce raw calcium as a by-product. Calcium happens to react with the CO_2 in rainwater to produce limestone, a solid that won't float up into the air and trap heat. So maybe we could open a giant ant farm in Siberia and let them have at it. As an alternative, we could exploit one-celled aquatic creatures called phytoplankton. Many regions in the ocean lack iron, an essential nutrient, so dumping powdered iron into those seas could produce huge "blooms" of phytoplankton. Phytoplankton build exoskeletons out of carbon, which they cull from the CO_2 in the air. When they die and sink, that carbon gets deposited at the bottom of the ocean. Now, ants and phytoplankton might seem too puny to tackle climate change, but we humans tend to underestimate just how many of these critters exist and just how much work they can do. As living creatures, they can also reproduce on their own, eliminating the need for humans to maintain them.

Another approach to climate engineering involves blocking sunlight before it reaches the ground and gets converted into heat. To this end some scientists have suggested launching giant mirrors into orbit or spraying seawater into clouds to make them whiter, fluffier, and more reflective. The most talked-about idea involves spraying megatons of sulfur dioxide into the stratosphere, since SO_2 also reflects sunlight back into space. (It's like an anti–greenhouse gas.) Although it once contributed to acid rain, this sulfur dioxide wouldn't get washed out of the sky very easily because we'd be spraying it above the heights at which rain clouds form. Another big benefit of sulfur dioxide is that, unlike with other approaches to climate engineering nature has already run some crude experiments for us: volcanoes often release this gas, and major eruptions such as Tambora and Krakatoa did cool the planet for several years. The downside is that, again, we could stumble into other problems that we didn't anticipate. If nothing else, SO_2 would dull the brilliant blue of our skies and obscure our view of the stars at night; sunsets, meanwhile, would look luridly red.

We could always combine several different tactics, of course— huge ant farms, supertankers spraying the seas with iron, cannons firing shells full of sulfur dioxide from atop Mount Kilimanjaro.

Unfortunately, once we start leaning on these technologies, we can never let up; we'll be bailing and bailing until kingdom come. Again, though, given the failure of human beings throughout history to ease up on consumption before everything collapses, engineering our way out of the mess seems like the most pragmatic option. And while this work will probably cost a lot—hundreds of billions of dollars per year—it actually seems cheap weighed against the prospect of extinguishing our entire frickin' species.

—◄○►—

But let's say that climate engineering fails and that Earth goes (pretty much literally) to hell. At that point, our only option might be starting over on another planet.

If we don't have much in the way of intergalactic star cruisers, we'd have to stay local. This would mean giving Mars or the moon an environmental makeover,* a process called terraforming. Some aspects of terraforming would involve simply tweaking what's already there. Martian soil, for instance, resembles the soil on Earth far more than its red color might suggest: it's chock-full of nutrients, and food crops that prefer alkaline soil would probably thrive there (provided they had air and didn't freeze, of course). The moon, however, would need more help to get its soil up to code. And both bodies would still need water and air. Luckily, we could fulfill both those needs by importing raw materials in the form of comets. Scientists have already landed probes on comets, and if one of those landers brought along an atomic bomb, the blast could nudge the comet off course and redirect it toward the target planet. Before the comet crash-landed we'd then smash it into smithereens with another nuke, after which its ice and gases and mineral nutrients would sprinkle down harmlessly onto the surface (at least in theory).

According to some estimates, just one hundred comets the size of

Halley's would transform the moon entirely, fertilizing its landscape and filling its "seas" like Tranquillty with actual water. (Mars, being larger, would require more comets.) Once we have water, getting a breathable atmosphere is straightforward: we simply import some oxygen-making algae and let them do their thing. (Some studies indicate that this process could take tens of thousands of years; others estimate far less time. Regardless, after we reach a certain minimum pressure, plants could be imported to speed things up.) As a bonus, we could bottle our most potent CFCs and ship them to Mars or the moon as well, where they'd actually do some good in warming those frigid bodies.

As Mars's atmosphere grew thicker, its sky—currently a creamy pink-yellow, due to dust—would start shading bluer and bluer as gases began to scatter sunlight. Something similar would happen to the moon's now black sky. And the moon itself would look different from down here on Earth. Some calculations suggest that a terraformed moon would glow five times brighter in our sky, with temperatures as balmy as Florida's. Given that climate, and the fact that its lower gravity would be easier on the joints, you could imagine the moon becoming a popular retirement spot.

Considering how much work it would take to transform an entire planet, though, the moon or Mars might not be the best long-term option. And in the really long term, transplanting ourselves within the solar system is no option at all, because the sun will eventually destroy everything around us. When the sun first blinked to life 4.5 billion years ago, it was 30 percent dimmer than now. It's been growing brighter and warmer ever since, and will probably grow warm enough in two billion years to boil away every ocean on Earth. Even if some sort of commando cockroach species survived that onslaught, nothing will survive the ultimate death of the sun roughly five billion years in the future, when it runs out of hydrogen fuel to burn. Several things will happen at that point, but the upshot is that the temperature of the core will rise significantly; as a result, creatures in this vicinity of the cosmos will once again learn the lesson, for the very last time, that gases expand as they get hotter. For the sun will quickly swell to more than 150 times its current diameter and metamorphose into a red giant star that, depending on whose calculation you credit, will either swallow Earth whole and vaporize it, or creep just close enough to give us a searing-hot kiss and reduce our beloved home to a cinder. Either way, Robert Frost guessed right: our world will end in fire.

Before that point, obviously, we'd have to colonize an exoplanet to survive. First things first, we'd have to figure out which exoplanets have air we can breathe, something we can check with telescopes and absorption spectrums. Next we'd have to build a gigantic spaceship to haul people to our new home. Thankfully, much of the raw material we'd need for star cruisers already exists in space, in the form of metallic asteroids that we could mine. Mining space rocks probably sounds ludicrous, but several space mining companies— some of them backed by Google and Microsoft billionaires— already exist and are already scouting for candidates among the tens of thousands of asteroids near Earth. Simple probes would act as mules, dragging the asteroids back toward Earth and parking them in gravitationally stable points in space where we can access them.

These space mining companies plan to make their initial profits on precious metals. An asteroid only five hundred yards across— one-twentieth the size of what killed the dinosaurs—could liberate more platinum than has ever been mined on Earth in its entire history. Later on, all the leftover iron in the asteroid could make nifty spaceships as well, ships we could actually build in space, as big as we wanted, since we wouldn't have to worry about lifting them off Earth's surface.

But the real prize on those asteroids might not be platinum or iron at all but the ice clinging to their surfaces. For some perspective here, the largest object in the asteroid belt, Ceres, appears to have more fresh water on it than all the lakes and rivers on Earth combined; most asteroids are smaller, but there's still a lot of water out there for the taking. Human beings traveling in space would need that water to drink, and splitting the H_2O could also produce hydrogen and oxygen to burn as fuel. Unlike cars, which move via friction, spaceships move by pushing little puffs of gas out their backsides and getting a boost of momentum—momentum they never lose to air resistance in the vacuum of space. (And if itty-bitty gas particles don't seem powerful enough for the task of moving a spaceship, well, all I can say is that you really haven't been paying attention this whole book…) Best of all, these ships could pick up more ice from other asteroids or comets along the way, using them like interstellar gas stations.

Splitting water molecules would also give the exonauts inside the spaceship extra oxygen to breathe. Most of their oxygen, though, would likely come from a much older gas-generating technology—plants, which they'd grow inside their cabins. The exonauts would also have to top off the cabins' internal atmosphere with nitrogen, both to keep the air pressure at Earth-like levels (imagine your ears being popped for decades at a time) and to mitigate the risk of fires, which burn wildly in pure oxygen. As for where to get this nitrogen, the crew would probably just grab some air from Earth before departing. In doing so, they'd inevitably suck up some of the argon and ammonia and other trace gases that make up our atmosphere, all of which would accompany us to our new home. Which seems fitting, given how much these gases have shaped our species, too.

As for which planet to strike out for, there are plenty of options, given that roughly 300 sextillion stars exist in our universe. (Put another way, it would take you several deep breaths to inhale that many molecules of air, and the number of air molecules you inhale with each breath is already gargantuan.) Statistically, the nearest habitable planet might be as close as twelve light-years away — a distance that, as life-spans increase, a human being could theoretically cover in one lifetime. During the journey, the people on the ship would need to exercise constantly to keep their bone densities and muscle masses high. (Arranging for parts of the ship to rotate like a slow centrifuge, creating artificial gravity, would help.) Beyond that chore, they'd occupy their days playing games and watching holographic movies and having babies and getting into arguments and doing every other thing that human beings do. Any astronomers aboard might get a kick out of watching the shapes of constellations morph (at least a little) as our position vis-à-vis the stars changed. Every so often, too, the spaceship might plow into a random pocket of space gas — the raw material of future solar systems.

Eventually our new home planet would appear — just a pixel at first, then a small smudge. At this point, scientists onboard would double-check that the planet really did have the atmospheric profile they assumed. Is there enough oxygen and ozone? Too much hydrogen sulfide or chlorine? What if there's loads of nitrous oxide, laughing gas — would we take one step outdoors and turn into blithering loons? We'd also have to search hard for any moons this planet had. Planets and their moons can have distinct atmospheres, with distinct gases on each. From afar, all these gases might blend together as a single absorption spectrum, because we wouldn't have the power to tease such small bodies apart. But up close we might find that, while some of these vital gases belong to the planet, some actually belong to its moon, which isn't much help.

Provided that everything checked out, we'd start picking out the planet's colors as we inched closer. Some shades would look familiar — oceans just as blue as Earth's, deserts just as tan. In contrast, depending on the peak output of our new sun, any forests of "plants" might look red or yellow rather than green. At long last we'd slip into orbit around this planet and would finally see the outlines of unfamiliar continents below. We'd have to be patient at this point: any planet worth calling home would have an atmosphere thick enough to burn our lumbering space condo into ash if we tried to land it. But a few brave souls could board a landing craft and make their way down. A few hours later they'd take their triumphant first steps on the new planet.

Still, for the long-term survival of our species, what happens next would be far more important. Depending on the ambient air pressure, the landing party might notice some odd things about this planet. If the atmosphere was much thicker than Earth's, any plantlike organisms would be shorter and more firmly anchored in the ground, to prevent them from toppling over in the stronger winds. Mountaintops would be warmer and easier to colonize in this case, since there'd be more air up there. And flying creatures could be substantially bigger, since generating lift would be easier. Indeed, the landing party would probably spend a few tense moments scanning the skies for predators as it stepped out. Finally, though, would come the moment they'd traveled trillions of miles for: when one member of the posse would nod at his comrades and start to unstrap his helmet.

That first sip of air might well bring death. Some trace gas— something we didn't even know we should worry about—might sear his lungs or paralyze his neurons. Much more likely, this strange air would burn his throat a little, not unlike a newborn's first breath. Things might smell funny, too, damp or dank or rotten. But there probably wouldn't be any reason to panic or gasp. He'd probably just laugh a little in relief, and take a few deep breaths to clear his lungs.

As he did so, something amazing would happen. All the nitrogen and other gases in his lungs, the air he'd carried with him from home, would trickle forth and escape. After all that distance, this tiny bit of air from his home planet would burst forth and consecrate the air of his new home. The atmosphere of Earth and this new planet would now be forever intertwined. The same thing would happen when the other exonauts removed their helmets and washed their lungs out, adding their own Earth-born molecules to the mix. And because the average person always carries within her lungs a molecule or two that Julius Caesar breathed during his final moments, several Caesar molecules would now pirouette upward and carry his story forth on this new planet.

There's no reason to limit ourselves to Caesar, either. As more and more people began descending from the mother ship and emptying their lungs, molecules that Harry Truman breathed at Mount Saint Helens; molecules that witnessed the atomic bomb blasts at Hiroshima and Bikini; molecules that mingled with the nitrous oxide in Humphry Davy's lungs and that swirled around Mount Everest while James Clerk Maxwell wondered what made the sky blue—all of them would join this new planet, too. As would a few molecules from your own life, the air coursing through your lungs during your first wail in the delivery room, your first kiss, your final breath years and years hence.

When we speak of endings, we say *dust to dust, ashes to ashes,* but that's not quite right—there's more to it. Every molecule in our bodies started off life as a gas, and long after our demise, when the big red bloated sun swallows everything around us, all those atoms will return to a gaseous state. A few lucky molecules could even get a second chance somewhere else. Some tiny bit of you—molecules that danced inside your body, maybe even that formed your body—could live on in a distant world. The idea of some part of me living on after I die sounds a lot like the stories of heaven I used to hear as a kid—except that here it's really true, it really will happen. We've talked all book about the millions and billions and septillions of stories swirling around us, coursing into and out of our lungs every second. You can capture the entire history of the world in a single breath. Journeying to another planet will inevitably, in some small way, keep those stories alive a little longer. Dust to dust, gases to gases.

Acknowledgments

Like the molecules in a breath, so many individual pieces had to come together to make this book possible, and I once again marvel at how generous everyone was with their help. A few words on a page aren't enough to express my gratitude, and if I've left anyone off this list, I remain thankful, if embarrassed.

For loved ones, I'd like to thank my dad for his love of science and of great lines, and my mom for her storytelling and for being a good sport. (I think I've teased her in every book so far.) Each year I feel a little luckier to know my siblings Ben and Becca, and it's been a delight to watch my little niece and nephew, Penny and Harry, become real people. So much has changed with my friends in Washington, D.C., and South Dakota and elsewhere, but through marriages and moves and everything else, we're all still sharing the good times.

Both my agent, Rick Broadhead, and my editor, John Parsley, saw how much potential this idea had, and helped me shape and refine the book throughout. *Caesar's Last Breath* wouldn't be here without them. I also want to thank everyone else in and around Little, Brown who worked with me on this book and others, including Malin von Euler-Hogan, Chris Jerome, Michael Noon, and Julie Ertl.

Finally, I offer a special thanks to the many, many brainy scientists and historians who contributed to individual chapters and passages, either by fleshing out stories, helping me hunt down information, or offering their time to explain something. They're too numerous to list here, but rest assured that I haven't forgotten your help...

Notes and Miscellanea

Welcome to the endnotes! Whenever you see an asterisk in the text (*), you can flip back here to find bonus material about the topic at hand. If you want to flip back immediately for each note, please do; or if you prefer, you can read all the notes at once after each chapter, like an epilogue. But do check back. There are some gems buried here, I promise...

Chapter One: Earth's Early Art

p. 40, something like 75,000 calories: This note's a bit of a trivia question: Where is most of the chemical energy in your body stored? Most people say in fat or muscle, but it's actually inside water molecules. Specifically, inside the O–H bonds that keep H_2O together: it would have taken an additional 550,000 food calories to snap all the O–H bonds in Harry Truman and break him down completely into individual atoms. (And again, older people have less water in them than younger people. For someone my age that would rise to around 670,000 calories.) As with the other caloric figures here, that's only an estimate: someone who makes different assumptions might come up with different figures. But it does give a rough idea of how much energy water molecules contain.

Interlude: The Exploding Lake

p. 47, lest they stir up something evil: Geologists do know of a few other places where clouds of carbon dioxide billow up from

The Grotta del Cane in Italy, where people once let small dogs pass out for sport.

time to time. There's Death Gulch in Yellowstone National Park, which has killed many an unwitting bird and rodent, even a few grizzlies. There's Grotta del Cane (Cave of the Dogs) in Italy, a popular tourist destination in the 1800s. The heavy CO_2 gas in the grotto hugs the floor, and visitors back then would amuse themselves by letting short dogs run around until they passed out. Finally, and most poignantly, there's Lake Monoun in Cameroon. It sits just sixty miles from Lake Nyos, and thirty-seven people died there under eerily similar circumstances in August 1984, suffocating overnight after a cloud of gas enveloped the region. A geologist who visited Monoun later that year tried to take samples from the bottom of the lake and found that the lids of his bottles kept popping off, the water was so fizzy. He sent a dispatch about the danger of these crater lakes to a few science journals, but they rejected the idea as far-fetched.

The most deadly gas outburst in history took place in Iceland in 1783, when a volcanic fissure spewed poisonous gas for eight months, ultimately releasing 7 million tons of hydrochloric acid, 15 million tons of hydrofluoric acid, and 122 million tons of sulfur dioxide. Locals called the event the *Moduhardindin,* or the "mist hardships," after the strange, noxious fumes that emerged—"air bitter as seaweed and reeking of rot," one witness remembered. The mists killed 80 percent of the sheep in Iceland, plus half the cattle and horses. Ten thousand people there also died—one-fifth of the population—mostly of starvation. When the mists wafted over to England, they mixed with water vapor to form sulfuric acid, killing twenty thousand more people. The mists also killed crops across huge swaths of Europe, inducing long-term food shortages that helped spark the French Revolution six years later.

Chapter Two: The Devil in the Air

p. 53, it has outlasted everything else that volcanoes spat out: Just to be clear: the nitrogen that began to accumulate in the air billions of years ago came mostly from volcanoes (either directly, or from the breakdown of volcanic ammonia). And much of that N_2 is still around today. But certain bacteria absorb and metabolize nitrogen, converting it into biologically useful products. Other bacteria then reverse that process and release nitrogen back into the air as N_2. So, while much of the nitrogen you're breathing right now came directly from volcanoes, some of it might have been reincarnated a few times in living things along the way.

p. 55, Chincha Islands, off Peru: Due to several meteorological quirks, it basically never rains on the Chincha Islands. (You've heard of those people raised in Florida or California who've never seen snow? People who live near the Chinchas—especially in the Atacama Desert

The mounds of guano on the Chincha Islands off Peru. The humans in the middle provide a sense of scale.

along the coast of Chile—can go a lifetime without seeing rain.) This lack of moisture makes the guano there extra potent, since rainwater tends to leach away precious nutrients as it seeps through guano deposits toward the ground.

By the 1850s the Chinchas were exporting millions of tons of guano per year, and workers there were suffering through some of the worst labor conditions human beings have ever endured. Most workers were kidnapped from China, Polynesia, or New Guinea; but after a few days on the islands you couldn't really tell their ethnicity anymore, since they were all caked in white guano dust. The utter lack of moisture left their lips, tongues, and noses cracked; some lacked even the tears to flush the ammonia fumes from their eyes. They spent twenty hours a day pounding the petrified bird poop with picks and scooping it up with shovels, and when their hands were too cracked to hold tools anymore, their masters lashed their forearms to wheelbarrows and bid them haul the guano to the cliffs on the islands' edge. There, they dumped the guano down chutes into barges waiting hundreds of feet below. After a few months, many workers flung themselves down the chutes to commit suicide rather than face another day of toil.

p. 56, guano greed a century before: Several people also used the Guano Islands Act to lay claim to islands that didn't exist—mirages that sailors had seen, spurious Isles of This or That on old maps. Oddly, Ernest Hemingway's brother Leicester invoked it in 1964 in founding the Republic of New Atlantis—a sovereign nation that consisted of nothing more than an eight-foot-by-thirty-foot bamboo raft anchored off Jamaica. Leicester was trying to establish territorial claims in the surrounding ocean to protect marine habitats, and he issued several denominations of stamps to raise money for his project.

p. 68, made the stalemate worse: Here's a bizarre fact: several gases currently banned in international warfare can nevertheless be used by police forces in the United States to subdue riots or other domestic unrest. To be sure, we aren't talking mustard gas or phosgene, mostly your nastier sorts of tear gas. Still, the U.S. government apparently believes that it's inhumane and cruel to use these gases against foreign combatants in a war but perfectly acceptable to turn them on its own people.

p. 73, Word of Haber's pathetic end: Before we leave Fritz Haber, I'd like to examine one more facet of his story—why his chemical warfare work seemed, and still seems, so barbaric. We now live in an age of Kalashnikovs and ICBMs, weapons capable of killing far more people far more quickly. Yet gas attacks still strike us as uniquely terrifying. Why?

First: unlike, say, most scientists involved with the Manhattan Project, Haber expressed no public pangs, no distress, no remorse for his role in gas warfare. What's more, the hornets that Haber kicked up during his lifetime continued to sting after his death. As mentioned, he cloaked his later gas warfare research under the guise of "insecticide" work. One of those insecticides, Zyklon A, was later tweaked into Zyklon B, the poison gas of choice for killing Jews—including some of Haber's relatives—at Auschwitz and Dachau and other Nazi camps.

Another reason gas attacks stir up such terror is that they threaten our basic biology in a way that machine guns and nuclear warheads cannot. I think a quick digression will help make this point clear. In my previous book, on neuroscience, I discussed a woman named S.M. who, because of brain damage, seemed incapable of feeling fear. Scientists drove her to exotic pet stores to handle snakes and tarantulas, and she didn't bat an eye. They ran her through haunted houses and showed her slasher films, and she shrugged. She even came close to dying several times—a mugger once held a knife to her throat in a park—and she remained unperturbed throughout. No pounding heart, no jolt of panic, no nothing. Scientists eventually concluded that she couldn't feel fear for anything.

It turns out that wasn't quite true. Just to see what would happen, S.M.'s doctors filled a tank with carbon dioxide–enriched air one day and asked S.M. to inhale it through a mask. Now, when you're held underwater, it's not the lack of oxygen that panics you, it's the buildup of CO_2. But given her lack of fear in every other context, her doctors suspected that S.M. would remain calm here. To their shock, she began hollering after a few hits of air and clawing at her mask, trying to tear it off her face. This one thing, a gas, could still frighten her. Scientists concluded from this and related work that human beings have a second, independent fear system lurking within the brain, one that closely monitors our air supply.

That, I think, is why Haber's work appalls us. When we can't breathe, we lose our minds, we start thrashing It's a biological, totally hardwired fear, and messing with our air supply will trip wires in the brain that bullets and other modern weapons simply can't. It's similar to the way that snakes and sharks terrify us far more than cars ever will, even though we're vastly more likely to die in a car wreck. Poisonous air belongs in the pantheon of primal fears.

Overall, I find Haber one of the most fascinating characters in science history. No one else so perfectly embodies the Faustian nature of science, its simultaneous promise and peril. A colleague of his once said of Haber, "He wanted to be both your best friend and God at the same time." He failed at both, and the disappointment we feel about Haber is all the more acute given his previous state of grace.

Interlude: Welding a Dangerous Weapon

p. 79, roughly similar chemically: To be sure, rusting and burning aren't identical: rusting usually requires water, for one thing, while water tends to snuff out flames. And both rusting and burning can produce several different types of iron oxides, depending on the circumstances. But the two processes do have a lot in common chemically: both involve oxygen atoms attacking iron and forming new compounds.

Chapter Three: The Curse and Blessing of Oxygen

p. 89, with everyday equipment: Priestley's amateurism shines through in other ways as well, beyond his crude equipment. For one thing, even in his scientific papers he often confesses how surprised he is at the results of his experiments. You get the feeling he walked around with his jaw open half the time, muttering and shaking his head in wonder. I love these admissions because they capture the very thing that draws most people toward science—the joy of discovering things about the natural world. And I can't help but think that students would learn a lot more about science from Priestley's style of honestly recording what he felt at each step than from the pristine, almost cold-blooded style that dominates scientific discourse today.

p. 93, they asked Lavoisier to investigate: Three years after his work with the French navy, Lavoisier joined the Régie de Poudres, which manufactured gunpowder for the military. Before this the Régie was a typical government bureau, beset with waste and sloth, but Lavoisier whipped his new charges into shape, and pretty soon France became self-sufficient in gunpowder for the first time. It even began exporting gunpowder to America. The United States likely wouldn't have won independence without this help, since Great Britain had cut the colonies off.

p. 98, anaerobic bacteria: When someone dies, bacteria begin to decompose the body. And although both aerobic and anaerobic bacteria contribute here, it's the anaerobic ones that produce the smelly gases we associate with rotten flesh. These include the aptly named molecules putrescine, $NH_2(CH_2)_4NH_2$, and cadaverine, $NH_2(CH_2)_5NH_2$.

p. 101, "inhale" through pores in their skin: It's easy to see how aboveground structures such as fruits, flowers, and woody stems can "inhale" oxygen. And the green parts of plants can of course just use the oxygen they're already making via photosynthesis. But how do the roots of plants "breathe"? Luckily, air can penetrate dirt easily: soil is pretty porous stuff, mostly because earthworms are constantly chewing through it and breaking it down. (Charles Darwin first discovered this fact about worms in a series of experiments.) This also explains why most plants can't survive in standing water—their roots starve from too little oxygen. In other words, plants die underwater for much the same reason that human beings do.

Chapter Four: The Wonder-Working Gas of Delight

p. 132, sodium: I can't resist passing along this clerihew about Davy's supposed regret over finding one of these elements:

Sir Humphry Davy
detested gravy.
He lived in the odium
of having discovered sodium.

Later in this book we'll also get into gaseous refrigeration, the inspiration for another grand clerihew, about low-temperature pioneer James Dewar:

Professor Dewar
Is better than you are.
None of you asses
Can condense gases.

p. 139, ether, a cheap high: One town in Ireland had such high rates of ether addiction that you could reportedly smell it from a half mile away. In addition to inhaling ether, people there often drank it with milk. Some set their mouths on fire when they smoked tobacco after drinking it.

p. 139, even a few fish: Given our similar lungs and nervous systems, I wasn't surprised to hear that anesthesia works on other animals. But it boggled my mind to learn that anesthesia works on some plants as well. You can put Venus flytraps under, for instance, and they won't snap their jaws shut when bugs land. This fact has sparked all sorts of philosophical disputes among botanists about whether plants have some sort of slow-motion consciousness or intelligence.

p. 143, rival claims popped up: Morton's biggest rival here was Charles Thomas Jackson, a manic New England doctor who may or may not have first suggested using ether as an anesthetic while in Morton's company. Jackson did have one major rhetorical advantage in this spat: his brother-in-law, Ralph Waldo Emerson, who championed Jackson's claims in public for decades. (One of Jackson's sisters, incidentally, first introduced Emerson and Henry David Thoreau.) Jackson also got tangled up in a dispute with Samuel Morse over the origins of the telegraph. It seems that Jackson had loads of revolutionary ideas, but he lacked either the gumption or guts to do more than just spout off about them.

Charles Thomas Jackson, brilliant but feckless inventor.

Interlude: Le Pétomane

p. 156, why don't we "speak" through our rear ends?: For a full discussion of why we don't talk out of our asses (at least not all the time), see Robert Provine's delightful book on bodily functions, *Curious Behavior.*

Chapter Five: Controlled Chaos

p. 158, Aristotle's dictum that nature abhors [a vacuum]: Oddly, quantum physics tells us that good old Ari might have been right after all, since even vacuums aren't totally empty: subatomic particles pop into and out of existence all the time inside them. Vacuums also have an intrinsic energy density, which means (per $E = mc^2$)

A device for automatically opening temple doors with steam power, invented by Hero of Alexandria in the first century.

that they contain mass. In fact, this energy density might be the mysterious "dark energy" that cosmologists think is causing the universe to expand.

p. 164, Human beings have long used water to drive machinery: The ancient Greeks invented several hydrological machines for keeping time and grinding grain, among other tasks. Hero of Alexandria in the second century AD even built steam-powered robots that sang and danced. (Really.) Priests used his devices in their temples as well, to close doors and move objects around on altars without anyone touching them. Not only was this prestidigitation fun, it awed the commoners, who assumed that the priests could summon gods or other spirits at will. The difference between Hero and Watt is that Hero seems to have limited himself to building toys, while Watt built steam engines to do physical labor. Nothing wrong with toys, of course—many technologies start off that way. But nobody built on Hero's work, either.

p. 166, no vacuum pump on Earth: Since vacuum pumps on Earth can't lift water past 34 feet, soda straws fail after that height, too: if you tried to suck up juice from atop a four-story building, you'd never get a drop.

But what about on Venus? The ambient air pressure on Venus is ninety times greater than on Earth, which means that the air there can push ninety times harder. And things are even better than that. Because Venus is slightly smaller than Earth, liquids traveling up the straw would feel less of a tug due to gravity (about 10 percent less). All in all, then, you could suck up water through a straw over 3,400 feet on Venus. Meanwhile, a straw could suck up water only 7 inches on Mars, since the air pressure there is so low (0.6 percent of Earth's air pressure). On the moon, which has virtually no atmosphere, a straw would raise water 6.5 trillionths of an inch.

p. 176, releases those gases almost instantly: Nitroglycerin explodes thousands of times faster than even an airbag does. Most airbags produce gas by sending a pulse of electricity through chemicals such as sodium azide (NaN_3), which decomposes into pure sodium and N_2 gas. At standard temperature and pressure, just 100 grams of NaN_3 can produce 50 liters of gas within 0.04 seconds. Sounds impressive, but 100 grams of nitro in those same conditions produces 70 liters in one ten-thousandth of the time.

Interlude: Steeling Yourself for Tragedy

p. 186, Shakespeare set *Macbeth*: Speaking of the Scottish play, before he became a ham-fisted poet, McGonagall had tried to make a living as a ham-fisted actor. (Indeed, he never wrote a poem in his life until age fifty-two.) Alas, he was no more competent at the performing arts than the poetic ones. He was most notorious for a turn he took onstage as Macbeth. Knowing what a disaster he would be, the theater manager reportedly made him pay for the honor of playing the part, and he didn't disappoint. When it came time for Macduff to slay him and end the tragedy, McGonagall refused to go down. In fact, he turned on the other actor with his sword and almost lopped his ear off. Macduff finally had to tackle him and drag him offstage.

p. 188, carbon monoxide (CO): Incidentally, carbon monoxide kills people because it bonds to iron even more readily than oxygen does. Inside your red blood cells there's a molecule called hemoglobin. Hemoglobin contains several atoms of iron at its core, and these iron atoms recruit and deliver oxygen to cells. But if there's carbon monoxide in your blood (from breathing it), the CO elbows the oxygen aside and latches onto hemoglobin instead. As a result, red blood cells can't deliver oxygen anymore. To make matters worse, carbon monoxide is really hard to break down: CO has the strongest bond in nature, a triple bond even stronger than N_2's triple.

Chapter Six: Into the Blue

p. 198, Montgolfier envisioned the world's first balloon: Some sources say that a different sequence of events inspired the hot-air balloon. In this case, Montgolfier was brooding over a newspaper clipping about the long-standing French siege of Gibraltar, a fort impenetrable by land or sea. In the middle of this reverie he supposedly looked up and saw some scraps of paper and ash wafting over that evening's fire. They almost seemed to be flying, and all at once he envisioned a way to attack Gibraltar from the air.

p. 202, the bubbles in the bottle...whooshed out: If you've ever dropped a can of pop or beer, someone might have advised you to tap the top or sides before opening it, to prevent the beverage from whooshing out. Here's why. When you jostle a carbonated beverage, carbon dioxide bubbles gather on the inner surface of the metal. Tapping the can knocks these bubbles loose and causes them to rise within the container and gather near the top. So when you open the can a moment later, the bubbles—because they're no longer submerged in the liquid—won't drag any liquid with them when they rush out. The beverage might taste flatter, but you'll keep your hand dry and the floor clean.

p. 216, the Greek word for "lazy": The word *argon* appears in the original New Testament, which was written in Greek. In the parable of the workers in the vineyard (Matthew 20:3), Jesus says, "And he went out about the third hour, and saw others standing idle [*argon*] in the marketplace."

By the by, the seven other chemical elements mentioned in the Bible are gold, silver, lead, iron, copper, tin, and sulfur.

p. 217, called his gas helium: The discovery of helium in the sun by a crippled French astronomer named Jules Janssen is an inspiring story. Janssen also has a link to balloon history, because he quite bravely—at the risk of getting shot as a spy if he didn't make it—escaped the German army's siege of Paris in 1870 in a rickety balloon, in order to observe an eclipse in Africa. Alas, the full saga won't fit into a note, but I have written something up for my website. See http://samkean.com/extras/clb-notes.html.

There are additional notes on other topics there. And if so moved, you can drop me a line as well—I love to hear from readers: http://samkean.com/samkean.php#contact.

p. 218, every known noble gas: I'm fudging a little in claiming that Ramsay contributed to discovering every noble gas, since this column on the periodic table now includes element 118 (oganesson), which wasn't discovered until 2006. Then again, because 118 is such a heavy element, it probably has a distorted electron structure and therefore might not behave like a noble gas. No one knows.

p. 219, cranky old Mendeleev smiled: Mendeleev learned his lesson a little too well, in fact, and started seeing noble gases where they didn't exist. Physics was facing a crisis around this time with regard to light waves. As far as scientists knew then, all waves needed a medium in order to propagate: tidal waves needed water, sound waves needed air, the Wave needed drunken soccer fans, and so on. By analogy, light waves supposedly needed a medium called the luminiferous æther. Problem was, no one had ever seen this æther. No one had detected it in experiments, either, despite decades of searching. So in a fit of enthusiasm, Mendeleev called on the periodic table to rescue physics. He proposed that æther was a very tiny, very fine noble gas called newtonium that pervaded all matter. And by tiny, he basically meant infinitesimal: he estimated the mass of newtonium as one ten-billionth of a hydrogen atom. A spirited debate ensued until 1905, when Einstein's theory of relativity obliterated the need for the luminiferous æther. But for a few years there, the noble gases seemed to explain the nature of light itself.

p. 221, Nitrogen and oxygen and argon alone: Okay, not *quite* alone. You see, Rayleigh ended up missing something that would have — should have — killed his theory. The problem stemmed from a property of waves called interference. Imagine two light rays about to collide. If they happen to be exactly out of sync — that is, if one has crests exactly where the other has valleys, and vice versa — then when they meet, they'll cancel each other out. That is, they'll destroy each other and leave nothing behind. And it turns out that in an atmosphere where the air is distributed evenly, any blue light that gets scattered would almost certainly meet its death on its journey toward the ground. As a result, all this blue light should be obliterated before it reaches our eyes.

What saves Rayleigh's explanation is that our atmosphere, despite being pretty darn near uniform, isn't quite: there are absurdly small fluctuations in density, and those are enough to salvage his explanation. In case you're wondering who pointed all this out and saved Rayleigh, it was some guy named Albert Einstein.

Chapter Seven: The Fallout of Fallout

p. 233, the goats had undergone conditioning: They chose goats for a reason. Back during World War I, psychologists interested in shell shock needed an animal model to study. So a former student of Ivan Pavlov opened a lab at Cornell University where he basically spent several years scaring the piss out of barnyard animals, to give them shell shock. (Reporters nicknamed it the "heebie-jeebies farm.") Rabbits proved too simple-minded for the work, since they rarely developed a complex about loud noises. Pigs and dogs proved too smart — their behavior was too complex, their reactions too varied. Goats proved just right. And the psychologist in charge did learn something important about shell shock. Namely, that it's not the explosions themselves that rattle people, nor the injuries they cause. Rather, it's the *anticipation* of them—the stress of thinking about them, hour after hour, night after night—that tears people up and breaks their spirit.

Regardless, the experiments with the psychotic goats on Bikini failed. As film footage later revealed, the goats barely blinked when the nuke went off. They had been eating hay beforehand, and they continued to munch afterward, unperturbed. (Incidentally, the footage also caught a rat giving birth at the moment of the explosion; scientists dubbed her three pups Alpha, Beta, and Gamma.)

p. 236, The Manhattan Project wasn't a scientific breakthrough: Don't take my word for it. Richard Feynman himself said, "All science stopped during the war, except for the little bit that was done at Los Alamos. And that was not much science. It was mostly engineering." So it's an interesting question why the physicists of the Manhattan Project still receive the lion's share of the credit, while the chemists and engineers remain anonymous. I think a few factors came into play. One, you had great characters among the physicists: the court jester Feynman, the "American Prometheus" Robert Oppenheimer, the foreign savant Enrico Fermi, the Soviet spy Karl Fuchs, and so on. Beyond Glenn Seaborg, the chemists lacked for big names, and Seaborg wasn't exactly charisma personified. Plus, physicists can be pretty chauvinistic about their domain, and it was a physicist who prepared the first official report on the bomb. It downplayed the role of chemists and thereby established a narrative that future historians followed. Finally, physicists had published most of their work on nuclear fission before the war; it was therefore public knowledge, and people could cite it freely in news stories. The chemical details for refining uranium and plutonium, meanwhile, remained classified.

p. 238, macho disdain: Here's an example from the 1930s. At a public lecture on the wonders of radioactivity, future Nobel laureate Ernest Lawrence brought out a vial of radioactive sodium as a prop.

Unfortunately it was so radioactive that it overwhelmed the Geiger counter on hand. So Lawrence prepared some salt water with the sodium and called his colleague Robert Oppenheimer up to the stage. Down the hatch it went, and a minute later Oppie wrapped his hand around the radiation detector. It chattered like a squirrel, and everyone present had a good laugh.

Typical 1950s fallout shelter. (Photo courtesy of the National Archives)

p. 249, among the survivors: In addition to testing how well buildings held up to nuclear blasts, the government also tested how well people held up when confined to fallout shelters for weeks at a time. Some people treated the confinement as a lark—one couple spent their honeymoon there—but most emerged in a grim state. One family took to drinking; even their three-year-old got a shot to shut him up. Another family used the ventilation shaft built into the structure as a reward. As you can imagine, going to the bathroom in a confined space created quite a stink, but whenever the children behaved themselves for a few hours, the parents allowed them to turn the crank that jettisoned the nasty gases into the outside world. Overall, while most people considered the fallout shelters an adventure at first, by day four they'd often sunk into a funk.

p. 251, Pop culture weighed in: Not all pop culture derided nuclear weapons. The first book illustrated by Maurice Sendak, later the author of *Where the Wild Things Are,* was called *Atomics for the Millions,* and it was generally positive. Not that Sendak really understood that. Sendak took on the project only because he had failed all his science classes in high school and needed at least one passing grade. So his science teacher, who'd authored the book, paid him $100 and gave him a boatload of extra credit for sketching out some anthropomorphic atoms undergoing fission. Sendak later called himself "the dumbest kid he ever had in his class.... He had to explain each picture."

p. 252, same amount of damage to your tissue: Determining the biological effects of radioactive particles is a messy business. Different elements release different particles at different rates, and some of those particles do more damage than others. (What's more, some particles are relatively harmless outside the body but will devastate your tissue if swallowed or inhaled.) To make things even trickier, different body tissues absorb different radioactive particles at different rates, and not everything you absorb does biological damage anyway. I've certainly oversimplified in this chapter by reducing everything to "chest X-rays." But the alternative, using twenty different units with footnotes to explain each one, seemed worse. My approach is imperfect, and I recognize that, but it does provide some basis for comparison.

p. 256, the Bravo test: Overall, between 1946 and 1958, the United States exploded 67 nuclear bombs on Bikini and the nearby Enewetak

Atoll, with a collective yield of 7,200 Hiroshima bombs. That works out to 1.6 Hiroshimas per day over twelve years. Incredibly, even after this barrage, the U.S. military kept promising the native Bikinians that they could return home any day now. The military called its resettlement program Project Hardy—Return of the Native (groan). They never implemented it.

Incidentally, Bravo was not the biggest atomic bomb ever detonated. That honor belongs to the Soviet Union's Tsar Bomba, which unleashed the power of 3,000 Hiroshimas on the remote Siberian island of Novaya Zemlya on October 30, 1961. Unlike sleek, modern nuclear weapons, this one weighed 60,000 pounds, and broke windows up to 560 miles away.

Interlude: Albert Einstein and the People's Fridge

p. 259, choked on his eggs: Because I know you're curious, Einstein ate fried or scrambled eggs most mornings for breakfast, along with toast or rolls. As for his other culinary habits, he reportedly ate so much honey that his domestic staff purchased it by the bucket. Other favorites at the Einstein table included egg drop soup, salmon, mayonnaise, cold cuts, asparagus, pork with sweet chestnuts, and strawberry meringue. He liked his meat quite well done. "I am not a tiger," he once told his cook.

p. 261, all six were eventually liquefied: Although liquefying these gases was a great achievement, some people went a little overboard in celebrating it. After the Swiss scientist Raoul Pictet liquefied air, a newspaper headline in Brooklyn declared "Pictet, Foremost of Savants, Calls the Liquid the Elixir of Life, and Declares It Will Banish Poverty from the Earth."

Oddly enough, the six "permanent gases" were all liquefied before helium, nature's most stubborn gas, which didn't get liquefied until 1908. (Helium, which boils at −458°F, wasn't included in the canonical six because it hadn't been discovered in the early 1800s.) Even today, it's not easy to keep helium cool. A malfunction at the Large Hadron Collider at CERN in 2008 allowed six tons of liquid helium to boil off, and it took a year to get everything working again, at a cost of tens of millions of dollars.

p. 266, switched to this chlorofluorocarbon: File this under unintended consequences. The United States banned the production of chlorofluorocarbon gases (CFCs) within its borders on December 31, 1995. But the government allowed businesses that relied on CFCs — such as auto shops, which used them to refill air-conditioning units — to keep using recycled CFCs or to buy them from abroad. In other words, the demand stayed constant while the supply decreased significantly. As any economics teacher could have predicted, the price skyrocketed. People even began smuggling CFCs into the United States to make a quick buck. Reportedly these gases cost around $2 per pound to make in China, India, and Russia but would fetch $20 per pound or more on the black market. By the late 1990s, ten thousand tons of illegal CFCs were being smuggled into the United States each year, mostly through Florida and Texas. In Miami, CFCs were the second most lucrative contraband on the black market, next to cocaine.

Chapter Eight: Weather Wars

p. 269, the Beagle's captain, Robert FitzRoy: The Beagle's sponsors didn't choose Charles Darwin as ship naturalist because they valued his acumen with plants and animals; rather, they valued Darwin's status as a cultured gentleman, someone who could provide Captain FitzRoy with good conversation on the voyage. This wasn't just for FitzRoy's comfort, either: the previous captain of the Beagle had gone stir-crazy because he had no one to talk to and had killed himself.

Robert FitzRoy.

In later years, as he ascended the ranks of the British navy, Fitz-Roy redoubled his efforts at predicting the weather, and in fact made quite a few enemies among British businessmen by forbidding small fishing boats to set out on days of predicted storms. (Fishermen of course hailed him as a hero.) Sadly, though, while Darwin kept the black dogs of depression from hounding FitzRoy during their voyage, FitzRoy couldn't keep them at bay forever, and he ended up killing himself in 1865. Ironically, Darwin probably contributed in some measure to his companion's death. FitzRoy was deeply religious, and he always felt guilty that someone under his command, on his ship, had unleashed the scourge of Darwinism on the world.

p. 277, magic of the ideal gas law: Thunder is another example of how gas laws influence the weather. Flashes of lightning cause a spike in the temperature of the surrounding air. This spike forces the air to expand in volume, which in turn creates a burst of noise. Lightning reaches such high temperatures, in fact—55,000°F, five times hotter than the surface of the sun—that the nearby air becomes a plasma, the same übergas that emerges during nuclear blasts.

p. 287, known as turbulence: Turbulence has a reputation as one of the most cussedly tricky topics in science. In 2000, the Clay Mathematics Institute offered a $1 million bounty to anyone who could make some headway in solving the equations that govern turbulence (the Navier-Stokes equations). No one has claimed the prize yet, and no one is expected to anytime soon. On his deathbed, in 1976, the quantum physicist Werner Heisenberg supposedly announced that when he met God, he was going to ask Him two questions: Why does relativity govern the large-scale structure of the universe? And why do fluids like air and water turn turbulent when they flow? "I really think," Heisenberg whispered, "He may have an answer to the first question."

Interlude: Rumbles from Roswell

p. 296, the noise stumbles into your ear: To be clear, the air molecules that leave her mouth are not themselves flying across the room and striking your eardrum. Sound isn't wind. Rather, each collision merely passes along the *energy* that those initial molecules had. (Of course, as with Caesar's last breath, some of those molecules from her mouth will reach you eventually, as they diffuse through the air. But not until long after the sound has dissipated.)

p. 299, the most intense noises: When Krakatoa blew its stack in 1883, the noise of the explosion destroyed the eardrums of several sailors well over a hundred miles from the blast, and the pressure waves created by the eruption continued to circle Earth for five days. The noise wasn't audible any longer, of course, but cities as distant as Toronto and London kept recording fluctuations in their barometers every thirty-four hours (thirty-four hours being the time it takes for sound to circle the globe).

p. 299, first worked out the physics of the sound channel in 1944: Ewing also discovered a sound channel in the ocean. Ocean water gets colder as you sink, and sound slows down in colder fluids. But the speed of sound in water also depends on other factors, like density and salinity, both of which increase with depth. The calculations get messy, but the upshot is that sound travels fastest in the topmost and bottommost layers of the ocean, and slowest at a depth of about 3,000 feet. As a result, sound waves in the ocean tend to bend toward that depth, drawn there as if by a magnet.

Ewing thought the sound channel in the ocean could help rescue pilots lost at sea. On every ocean-crossing flight, a pilot would carry with him a ping-pong-sized metal sphere in case of a crash landing. When dropped into the ocean, the sphere would sink, and it was thick enough to withstand crumpling until a depth of 3,000 feet. At this point, however, it imploded with a crunch, a noise about as loud as a firecracker's bang. That might not sound like much, but again, the sound channel effectively magnifies noises, and special sound buoys (with microphones dangling 3,000 feet below the waves) could still hear it. By triangulating the signal between several buoys, rescue teams could determine the pilot's coordinates.

Incidentally, many biologists believe that humpback whales take advantage of the sound channel to croon to their brethren over thousands of miles.

Chapter Nine: Putting on Alien Airs

p. 308, damn people to hell: Speaking of hell, get a load of this. According to Revelation 21:8, hell has a lake of brimstone, which is basically molten (i.e., liquid) sulfur. Sulfur remains a liquid only up until 832°F, which means that hell is theoretically colder than Venus.

It gets better. The Bible also says (Isaiah 30:26) that the moon in heaven shines as bright as the sun. And, depending on which translation you use, the sun itself seems to shine "sevenfold as the light of seven days." In other words, heaven has the equivalent of fifty suns: one moon-sun plus forty-nine sun-suns. Since the ambient temperature of a planet rises quickly with an increase of sunlight (it scales up according to the fourth power), the temperature of heaven in this interpretation probably reaches close to 1,000°F—making heaven hotter than hell!

I can't take credit for this gem. You can read a fuller treatment in "When Hell Freezes Over," by Ron DeLorenzo, in *Journal of Chemical Education*, volume 76 (1999), page 503.

p. 311, discover helium in the sun in 1868: You didn't think you'd escape this book without another discussion of intestinal gas, did you? According to the Internet, Jules Janssen—one of the astronomers who discovered helium in the sun in 1868—supposedly experimented on the family dog with helium, forcing him to breathe it and even giving le pooch a few helium enemas. (I leave it as an exercise for the reader to determine whether that would make his farts squeak.) But considering that Ramsay didn't isolate this gas on Earth until 1895, and then in only minute quantities, the story seems unlikely.

Incidentally, the doctors on the Apollo space program spent a great deal of effort monitoring the flatulence of astronauts. They did so partly out of curiosity, since they didn't know how zero gravity would affect digestion, and partly out of fear, since they didn't know whether internal pockets of gas would tear holes in the astronauts' abdomens in the low-pressure environment of space. Turns out they didn't need to worry. People who live in lower air pressure—including mountainous regions on Earth—do fart more readily. (Mountaineers sometimes speak of encounters with "Rocky Mountain barking spiders.") But they don't fart with nearly enough violence to harm themselves.

p. 314, highfalutin spiritual questions: One highfalutin question involves religious belief: would the discovery of extraterrestrial intelligence undermine people's faith in God and/or the afterlife?

Depends on their creed. Of the world's major religions, Hinduism and Buddhism seem to be in the best theological shape with regard to aliens, since both actively support the idea of life on other planets. Similarly, the Koran suggests that intelligent beings exist elsewhere, although it's not clear whether they would have to follow Islam. Judaism seems to treat aliens as more or less irrelevant.

The religion that would probably be thrown into the most chaos (aside from a few alien-friendly sects, like Mormonism) would be Christianity. Although some Catholic scholars support the idea of aliens, most denominations don't, especially evangelical and fundamentalist branches. The idea in fact introduces some pretty thorny theological problems about original sin and whether every intelligent being across the universe deserves salvation and a trip to heaven. Perhaps other intelligent beings get to go to heaven automatically— but that makes things a little unfair here on Earth, where people have to earn their way.

p. 315, a whopping 900°F: You might have noticed something funny about the numbers here. Earlier I gave the temperature of Venus as 860°F. So if you subtract the extra 900 degrees that Venus gets from greenhouse gases, you'd end up at −40°F. I also claimed that Earth's temperature, without greenhouse gases, would be around 0°F. But how can that be, if Venus lies an average of 26 million miles closer to the sun? The answer is that Venus is covered with fluffy white clouds of sulfuric acid that reflect sunlight back into space and lower the planet's temperature. Earth lacks sulfuric acid clouds and therefore would remain warmer. But the overall point stands: Earth gets dozens of extra degrees from greenhouse gases, Venus gets hundreds.

p. 317, more evaporation, and so on: Some caveats: The evaporation feedback loop, by making more water vapor, will also create more clouds, and clouds tend to reflect sunlight back into space and slightly cool the planet. On the other hand, warm air also warms the ocean, and even slightly warmer ocean water can speed the melting of certain ice formations that contain methane, which heats up the planet when it escapes. So the overall effect is complicated. Indeed, it's complications and secondary effects like this that make our climate so bloody hard to model. Changing one variable almost always affects a dozen other things.

p. 320, giving Mars or the moon an environmental makeover: You might think that we could colonize Mars or the moon by erecting a big geodesic dome and living inside it. Well, in the early 1990s eight scientists sealed themselves inside a biosphere in Arizona to test this idea. It didn't go well. They made it for a few years, but over that time span the oxygen inside the dome dropped from normal Earth concentrations (21 percent) to just 17 percent, the point at which human beings struggle to breathe. Somehow, 60,000 pounds of oxygen went missing, probably into the maws of bacteria in the soil. Carbon dioxide levels also fluctuated, since concrete structures within the dome tended to absorb CO_2. The overall lesson was that gases are pretty hard to hold on to even inside sealed containers. Importing comets might actually be the easier solution!

Works Cited

General

Allaby, Michael. *Atmosphere: A Scientific History of Air, Weather, and Climate.* New York: Facts on File, 2009.

Almqvist, Ebbe. *History of Industrial Gases.* New York: Kluwer Academic/Plenum Publishers, 2003.

Canfield, Donald E. *Oxygen: A Four Billion Year History.* Princeton, NJ: Princeton University Press, 2014.

Fenster, Julie. *Ether Day.* New York: Harper Perennial, 2002.

Fisher, David. *Much Ado about (Practically) Nothing: A History of the Noble Gases.* New York: Oxford University Press, 2010.

Greenberg, Arthur. *From Alchemy to Chemistry in Picture and Story.* Hoboken, NJ: John Wiley & Sons, 2007.

Hazen, Robert M. *The Story of Earth: The First 4.5 Billion Years, from Stardust to Living Planet.* New York: Viking, 2012.

Jay, Mike. *The Atmosphere of Heaven.* New Haven, CT: Yale University Press, 2000.

Introduction: The Last Breath

Dando-Collins, Stephen. *The Ides: Caesar's Murder and the War for Rome.* New York: John Wiley & Sons, 2010.

Goldsworthy, Adrian. *Caesar: Life of a Colossus.* New Haven, CT: Yale University Press, 2008.

Parenti, Michael. *The Assassination of Julius Caesar.* New York: The New Press, 2003.

Chapter One: Earth's Early Air

Carson, Rob. *Mount St. Helens*. Seattle, WA: Sasquatch Books, 2000.

Findley, Rowe. "Mountain with a Death Wish," *National Geographic* 159, no. 1 (1981): 3–33.

Mastrolorenzo, Giuseppe et al. "Lethal Thermal Impact at Periphery of Pyroclastic Surges: Evidences at Pompeii," *PLoS ONE* 5, no. 6 (June 2010): 1–12, http://journals.plos.org/plosone/article?id=10.1371/journal .pone.0011127.

Rosen, Shirley. *Truman of St. Helens*. Seattle, WA: Madrona Publishers, 1981.

Stylianidis, Nearchos, Olorunfunmi Adefioye-Giwa, and Zane Thornley. "Complete Vaporisation of a Human Body," *Journal of Interdisciplinary Science Topics* 2, no. 1 (2013): 1–4.

————. "Human Body Vaporisation," *Journal of Interdisciplinary Science Topics* 2, no. 1 (2013): 1–3.

Zahnle, Kevin, Laura Schaefer, and Bruce Fegley. "Earth's Earliest Atmospheres," *Cold Spring Harbor Perspectives in Biology* 2, no. 10 (October 2010): 1–17.

Interlude: The Exploding Lake

Baxter, Peter J., M. Kapila, and D. Mfonfu. "Lake Nyos Disaster, Cameroon, 1986," *British Medical Journal* 298, no. 5 (1989): 1437–1441.

Kling, George W. "The 1986 Lake Nyos Gas Disaster in Cameroon, West Africa," *Science* 236, no. 4798 (1987): 169–175.

Krajick, Kevin. "Defusing Africa's Killer Lakes," *Smithsonian* 34, no. 6 (2003): 46–50.

Scarth, Alwyn. *Vulcan's Fury*. New Haven, CT: Yale University Press, 1999.

Chapter Two: The Devil in the Air

Bown, Stephen. *A Most Damnable Invention*. New York: Thomas Dunne Books, 2005.

Craig, Peter. "Mankind in Peace, the Fatherland in War," *New Scientist* 101, no. 1395 (1984): 15–17.

Hager, Thomas. *The Alchemy of Air*. New York: Crown, 2008.

Interlude: Welding a Dangerous Weapon

Almqvist, Ebbe. *History of Industrial Gases*. New York: Kluwer Academic/ Plenum Publishers, 2003.

Chapter Three: The Curse and Blessing of Oxygen

Bell, Madison Smartt. *Lavoisier in the Year One*. New York: W. W. Norton, 2010.
Bygrave, Stephen. "'I Predict a Riot': Joseph Priestley and Languages of Enlightenment in Birmingham in 1791," *Romanticism* 18, no. 1 (2012): 70–88.
Malone, John. *It Doesn't Take a Rocket Scientist*. New York: John Wiley & Sons, 2002.
Poirier, Jean-Pierre, and Rebecca Balinski. *Lavoisier: Chemist, Biologist, Economist*. Philadelphia: University of Pennsylvania Press, 1998.
Rose, R. B. "The Priestley Riots of 1791," *Past & Present* 18, no. 1 (1960): 68–88.

Interlude: Hotter than the Dickens

Haight, Gordon. "Dickens and Lewes on Spontaneous Combustion," *Nineteenth-Century Fiction* 10, no. 1 (1955): 53–63.
Perkins, George. "Death by Spontaneous Combustion," *Dickensian* 60, no. 342 (1964): 57.
West, John. "Spontaneous Combustion, Dickens, Lewes, and Lavoisier," *Physiology* 9, no. 6 (1994): 276–278.

Chapter Four: The Wonder-Working Gas of Delight

Fenster, Julie. *Ether Day*. New York: Harper Perennial, 2002.
Holmes, Richard. *The Age of Wonder*. New York: Pantheon, 2009.
Jay, Mike. *The Atmosphere of Heaven*. New Haven, CT: Yale University Press, 2009.

Interlude: Le Pétomane

Moore, Alison. "The Spectacular Anus of Joseph Pujol," *French Cultural Studies* 24, no. 1 (2013): 27–43.
Nohain, Jean, and F. Caradec. *Le Pétomane*. London: Souvenir Press, 1992.

Provine, Robert. *Curious Behavior.* Cambridge, MA: Harvard University Press, 2012.

Suarez, F. L., J. Springfield, and M. D. Levitt. "Identification of Gases Responsible for the Odour of Human Flatus and Evaluation of a Device Purported to Reduce this Odour," *Gut* 43, no. 1 (1998): 100–104.

Suarez, F. L., J. K. Furne, J. Springfield, and M. D. Levitt. "Morning Breath Odor: Influence of Treatments on Sulfur Gases," *Journal of Dental Research* 79, no. 10 (2000): 1773–1777.

Chapter Five: Controlled Chaos

Bown, Stephen. *A Most Damnable Invention.* New York: Thomas Dunne Books, 2005.

Marsden, Ben. *Watt's Perfect Engine.* New York: Columbia University Press, 2004.

Interlude: Steeling Yourself for Tragedy

Bessemer, Henry. *Sir Henry Bessemer, F.R.S.: An Autobiography.* London: Offices of Engineering, 1905.

Jeans, William. *The Creators of the Age of Steel.* London: Chapman and Hall, 1973.

Lewis, Peter, and Ken Reynolds. "Forensic Engineering: A Reappraisal of the Tay Bridge Disaster," *Interdisciplinary Science Reviews* 27, no. 4 (2002): 287–298.

Chapter Six: Into the Blue

Fisher, David. *Much Ado about (Practically) Nothing: A History of the Noble Gases.* New York: Oxford University Press, 2010.

Fontani, Marco, Mariagrazia Costa, and Mary Virginia Orna. *The Lost Elements.* New York: Oxford University Press, 2014.

Gillispie, Charles Coulston. *The Montgolfier Brothers and the Invention of Aviation.* Princeton, NJ: Princeton University Press, 1983.

Holmes, Richard. *Falling Upwards: How We Took to the Air.* New York: Pantheon, 2013.

Wolfenden, John. "The Noble Gases and the Periodic Table," *Journal of Chemical Education* 46, no. 9 (1969): 569–575.

Interlude: Night Lights

Dewdney, Christopher. *Acquainted with the Night.* New York: Bloomsbury USA, 2008.

Ekirch, A. Roger. *At Day's Close: Night in Times Past.* New York: W. W. Norton, 2006.

Tomory, Leslie. *Progressive Enlightenment.* Cambridge, MA: MIT Press, 2012.

Chapter Seven: The Fallout of Fallout

Boese, Alex. *Electrified Sheep.* New York: Thomas Dunne Books, 2012.

Mahaffey, James. *Atomic Accidents.* New York: Pegasus, 2015.

National Park Service, "The Archaeology of the Atomic Bomb," Chapters One, Two, Three, and Five. http://www.nps.gov/parkhistory/online _books/swcrc/37/contents.htm. Accessed November 4, 2015.

Rhodes, Richard. *The Making of the Atomic Bomb.* New York: Simon & Schuster, 1988.

Simon, Steven, André Bouville, and Charles Land. "Fallout from Nuclear Weapons Tests and Cancer Risks," *New Scientist* 94, no. 1 (2006): 48–56.

Smith-Norris, Martha. " 'Only as Dust in the Face of the Wind': An Analysis of the BRAVO Nuclear Incident in the Pacific, 1954," *Journal of American–East Asian Relations* 6, no. 1 (1997): 1–34.

Welsome, Eileen. *The Plutonium Files.* New York: Dial Press, 1999.

Winkler, Allan. *Life Under a Cloud.* Champaign: University of Illinois Press, 1999.

Interlude: Albert Einstein and the People's Fridge

Bryson, Bill. *A Short History of Nearly Everything.* New York: Broadway Books, 2004.

Dannen, Gene. "The Einstein-Szilard Refrigerators," *Scientific American* 276, no. 1 (1997): 90–95.

Illy, József. *The Practical Einstein.* Baltimore: Johns Hopkins University Press, 2013.

Trainer, Matthew. "Albert Einstein's Patents," *World Patent Information* no. 2 (2006): 159–165.

Chapter Eight: Weather Wars

Fleming, James R. "The Climate Engineers," *Wilson Quarterly* (Spring 2007): 46–60.

Gleick, James. *Chaos*. New York: Penguin, 1987.

Langmuir, Irving. "Control of Precipitation from Cumulus Clouds by Various Seeding Techniques," *Science* 112, no. 2898 (1950): 35–41.

Lorenz, Edward. *The Essence of Chaos*. Seattle: University of Washington Press, 1995.

Interlude: Rumbles from Roswell

McAndrew, James. "The Roswell Report," Air Force Historical Studies Division, 1995. http://www.afhso.af.mil/shared/media/document/AFD-101027-030 .pdf. Accessed November 4, 2015.

Muller, Richard. *Physics for Future Presidents*. New York: W. W. Norton, 2009.

Pretor-Pinney, Gavin. *The Wave Watcher's Companion*. New York: Penguin, 2011.

Chapter Nine: Putting on Alien Airs

Bennett, Jeffrey. *Beyond UFOs*. Princeton, NJ: Princeton University Press, 2010.

Kasting, James, and David Catling. "Evolution of a Habitable Planet," *Annual Review of Astronomy and Astrophysics* 41, no.1 (2003): 429–463.

Index

Note: Italic page numbers refer to illustrations.

absorption refrigeration, 263–65
absorption spectrums, 311, 322, 324
acetylene (C_2H_2), 223, *223*, 226
acid rain, 317–19
acoustics
  and altitude, 203, 297
  and argon, 215–16
  and refraction, 297
  and sound as energy, 351n
  and sound channels, 296–300, 351n,
    351–52n
aerobic bacteria, 100–101
air. *See* atmosphere, terrestrial;
  atmospheres, extraterrestrial;
  gases
air pressure
  in Earth's early atmosphere, 25–26
  and flatulence, 353n
  and meteorology, 271, 276–77, 286
  and power of vacuum, 159–65, 167
  on Venus, 308, 340n
alcohol, 111, 114, 116–17, 139
alien life-forms, 304–09, 312–14,
  353–54n
ammonia (NH_3)
  absorption spectrum of, 311
  in balloons, 205
  and coal gas, 226
  discovery of, 89
  on exoplanets, 312
  in explosives, 63
  and extraterrestrial life, 309
  as fertilizer, 53–55, 59–60, 62, 69, 74
  Fritz Haber's efforts to create, 50–53,
    51, 56–60, 62, 69

  molecular atmospheric dispersal of,
    49, *49*
  and purification of nitrogen, 212
  and refrigeration, 262
  and space colonization, 323
  using nitrogen to create, 50–54,
    56–62
anaerobic bacteria, 98, 336n
anesthesia
  chloroform as, 144–47
  and consciousness, 148–50
  ether as, 138–47, *141*, 338n
  nitrous oxide as, 133–38, 142, 146
  physiological effects of, 147–49, 337n
animal subjects of nuclear tests, 233–35,
  247, 249, 344–45n
Archimedes, 203–06
argon (Ar)
  biblical references to, 342n
  compared to water vapor, 270
  discovery of, 213, 215–19
  in extraterrestrial atmospheres, 213,
    312
  molecular atmospheric dispersal of,
    196, *196*, 216, 252, 270, 312, 323
  and sky's blue color, 221, 344n
  and space colonization, 323
Aristotle, 158, 339n
ash, from volcanoes, 35, 37
asteroids, 26–28, 307, 322–23
atmosphere, terrestrial
  acoustics of, 203, 296–300, 351n
  and chlorofluorocarbons, 266,
    349–50n
  and climate engineering, 318–20

atmosphere, terrestrial *(cont.)*
  and cosmic rays, 252–53
  evolution of, 13, 15, 24–26, 28, 42, 48,
    53–54, 75, 82–83, 98–103, 119, 307,
    315–16, 331
  human-caused changes in, 229–30, 304
  layers of, 208
  molecular dispersal of gases in, 3–4,
    7–11, 17, 43, 49, 75, 111, 121, 151,
    158, 185, 196, 223, 231, 259, 268, 293
  and nuclear fallout, 250
  and sky's blue color, 221–22, 319, 326,
    344n
  sound channel of, 296–300, 351n
  and space colonization, 323, 325–26
  speed of molecules in, 23, 25
  trace gases in, 14
atmospheres, extraterrestrial
  and alien life-forms, 304–05
  and argon, 213, 312
  on exoplanets, 308–14, *310*, 324–26
  on Mars, 305–07, 315, 320–21, 340n,
    355n
  and space colonization, 13, 323–26
  on Venus, 305, 307–08, 315, 317,
    340n, 352n, 354n
atomic bombs. *See* nuclear weapons

bacteria
  and decomposition, 336n
  and flatulence, 155
  nitrogen-fixing, 54–55, 99, 331n
  and oxygen, 98–101, 103–04, 355n
  and rain, 275
Baker nuclear test, 246–47, *246*, 255–56
balloons
  and buoyancy of fluids, 205
  dangers of flying in, 202–03
  and development of gas laws, 206–08
  and Jules Janssen, 342–43n
  and meteorology, 286, 293–96, *294*
  Montgolfier brothers experiments with,
    197–99, *197*, 201, 206, 211, 342n
  Project Mogul, 300–303, *301*
  Robert brothers and Jacques Charles's
    experiments with, 199–202, *200*,
    206

scientific experiments using flight of,
    203, 206–08, 222
  and sound refraction, 297
BASF, 59, 61–62, 67, 69–71, 74
Beddoes, Thomas
  cow-house therapy of, 129–31, *130*
  and nitrous oxide, 122, *123*, 127–28,
    131–32, 134
  and Pneumatic Institution, 124–33
  reputation of, 121–22, 131–33
  and study of gases, 122–23, 131–32,
    150, 191
beryllium, 237, 243
Bessemer, Henry, 189–91, *189*, 194, 316
Big Thwack, 27, 100
Bikini Atoll, 232–33, *246*, 248, 255–57,
    326, 344–45n, 347–48n
Black, Joseph, 84–85, 123, 168
blasting caps, 178–79
Bleak House (Dickens), 112–15, *113*, 117
blood circulation, 115–16
Bosch, Carl, 50, 59–62, *59*, 66, 69–74, 109
Bouch, Thomas, 193–94
Boulton, Matthew, 170, 172, 174
Boyle, Robert, 84, 206
Brazel, Mac, 293–96, 302
breathing. *See* respiration
Brutus (Roman senator), 6–7, 10
butterfly effect, 290–91

Caesar, Julius, 4–12, *6*, 41, 326, 351n
calcium, 132, 251, 318
calcium hydroxyapatite, 40
Camuccini, Vincenzo, *6*
cancer, 249, 253–55
candles, 224–26
carbon, 23, 175, 226
carbon-14, 253–54
carbon dioxide (CO_2)
  and Arizona biosphere experiment,
    355n
  in carbonated beverages, 88, 342n
  and climate engineering, 318–19
  discovery of, 84–85, 89, 123, 168
  dissolved in water, 26
  erupting from lakes and caves, 43–48,
    45, 329–31n, *330*

escaping from magma, 25
and evolution of Earth's atmosphere,
315–16
on exoplanets, 312
and flatulence, 155
frozen, 268, 273, 275, 277–78
heat absorption by, 315–18
on Mars, 315
molecular atmospheric dispersal of,
43, *43*
and plants, 101, 104
and plate tectonics, 30
and purification of nitrogen, 212
and radioactivity, 253
reactivity with argon, 216
and respiration, 94–95
and Venus's atmosphere, 307–08, 315
carbon monoxide (CO), 126, 185,
187–88, 194–95, 260, 341–42n
Carter, Jimmy, 37–38
cast iron, 187–92, 194. *See also* iron
Castle Bravo nuclear test, 256–57,
347–48n
catalysts, 57, 60
Cavendish, Henry, 85–86, 96–97, 199,
213–14
chaos theory, 222, 291–92, 317
Charles, Jacques, 202, 206, 222
chemical weapons, 63–69, *65*, 73, 102,
157, 250, 333n
chest X-rays, 241, 244, 252–53, 257, 347n
Chincha Islands, 55–56, 331–33n, *332*
chlorine, 64–67, 210, 216, 266, 311
chlorofluorocarbons, 266, 313–14, 317,
321, 349–50n
chloroform, 144–47
Claude, Georges, 227–28
climate change, 316–19
climate engineering, 318–20
clouds
and climate engineering, 319
and fallout, 238, 248, *248*, 250, 255,
257, 300
seeding, 268, 270, 273–75, 277,
279–84, 292
on Venus, 305, 315n
coal gas, 224–26

coal liquefaction, 71–72
coke, 188
Coleridge, Samuel Taylor, 128, 132, 134
comets, 26, 320–21, 355n
computers, and meteorology, 284–91,
285, 293
consciousness, and anesthesia, 148–50
conservation of mass law, 97, 99, 104
continental drift, 29–30
copper, 212
cosmic rays, 252–53
cow-house therapy, 129–31, *130*
crop rotation, 55
cyanobacteria, 98–99

Daghlian, Harry, 239–45, 249
Darwin, Charles, 133, 269, 336n, 350n
Darwin, Erasmus, 105, 170, 174
Davy, Humphry, 125–33, *126*, *130*, 150,
209, 226, 326, 337n
Deinococcus radiodurans, 309
dentistry, 135–37, 139, 146
Dichlorodifluoromethane (CCl_2F_2), 259,
259
Dickens, Charles, 112–17, *113*
digestion, 155–56
dimethyl sulfide (C_2H_6S), 151, *151*, 156
DNA
criticality of nitrogen, 54, 62
and extraterrestrial life, 306, 309
and free radicals, 99
and radioactivity, 253, 255
and ultraviolet light, 297, 303
dry ice, 268, 273, 275, 277–78
Dynamite, 180–82, 238

earthquakes, 30–35
Earth's atmosphere. *See* atmosphere,
terrestrial
Einstein, Albert
diet of, 259, 348n
and Fritz Haber, 63, 73
and Adolph Hitler, 72
and refrigeration, 12, 259–60,
263–66, *264*
and sky's blue color, 344n
theory of relativity, 266–67, 343n

emission spectrums, 311
environmental warfare, 283
Espy, James, 270, 274, 281
ethanol (C_2H_5OH), 111, *111*
ether, 138–47, *141*, 337n, 338n
evaporation, and greenhouse gases,
    317–18, 354–55n
Ewing, Maurice, 296, 299–302, 351–52n
exoplanets, 308–14, *310*, 322, 324–26
explosives
    and ammonia, 63
    Dynamite, 180–82, 238
    gunpowder, 175, 177–78, 181, 336n
    and the Industrial Revolution, 184
    nitroglycerin, 115, 175–81, 183, 341n
extraterrestrial life, 304–09, 312–14,
    353–54n

fallout. *See* nuclear weapons
Ferdinand III (Holy Roman Emperor),
    159–60
Fermi, Enrico, 243, 345n
fertilizer, 53–56, 59–60, 62, 69, 74,
    331–33n, *332*, 333n
FitzRoy, Robert, 269, *349*, 350n
flatulence, 13, 116, 152–57, 338–39n, 353n
Fletcher, Thomas, 78–80
flight. *See* human flight
fluids, 71, 115, 164, 204–05, 287, 351n,
    351–52n
Franklin, Benjamin, 88, 202, 305
free radicals, 99, 266
frontal systems, 271, 287

gases. *See also* steam; *and specific gases*
    atmospheric molecular distribution
        of, 3–4, 7–11, 17, 43, 49, 75, 111,
        121, 151, 158, 185, 196, 223, 231,
        259, 268, 293
    compared to liquids and solids,
        23–24, 71, 168, 262
    development of gas theories, 82–86
    etymology of word, 83, 292
    on exoplanets, 310–14, *310*, 324
    and explosives, 175–84
    and formation of solar system, 23
    and history of Earth, 11–13

human harnessing of, 13
    ideal law of, 206–10, 292, 350n
    intestinal, 13, 116, 152–57, 338–39n,
        353n
    and lighting, 13, 223–28, 310–12
    liquefaction of, 260–62, 348–49n
    and meteorology, 269–70, 277, 350n
    and refrigeration, 60, 259, 266
    speed of gas molecules, 25
    John Strutt's study of, 211–17, 219–21,
        344n
    temperature/pressure/volume
        relationships in, 24, 26–27, 31, 152,
        206–10, 292, 322
    transformation of, 24, 26–27
    and vacuum, 159–63
    and volatility of liquids, 138
    volcanic origins of, 13, 15, 24–26, 28,
        31, 42, 48, 50, 53–54, 307
    in warfare, 63–69, *65*, 73, 102, 157,
        250, 333n
Gay-Lussac, Joseph-Louis, 206–08, 222
George III (king of England), 107
Germany, 53, 56, 59, 63–74, 219, 267
Gilda/Able nuclear test, 234, 246–47, 255
global warming, 318
gold, 25
Graves, Alvin, 242–45, 256–57
Graves, Elizabeth, 244, 256
gravity
    and hydrogen balloons, 205
    and rain-making, 274
    and solar system formation, 23, 27
    and vacuum-pumping water, 165–66,
        340n
Great Oxygenation Event, 100
greenhouse gases, 101, 308, 314–19,
    354n. *See also specific gases*
guano, 55–56, 331–33n, *332*, 333n
von Guericke, Otto, 158–64, *162*, 172
Guinness Brewing Co., 60, 261
gunpowder, 175, 177–78, 181, 336n

Haber, Clara, 63–64, 67, 73
Haber, Fritz
    and ammonia/nitrogen production,
        50–53, *51*, 56–60, 62, 69, 70, 74

death of, 73–74
and Albert Einstein, 63, 73
and gas warfare, 63–69, 73, 102, 250, 333–35n
and Adolph Hitler, 71–73
heart disease, and nitroglycerin, 183
heat. *See* temperature
Heisenberg, Werner, 351n
helium (He)
in balloons, 205
discovery of, 217–18, 311, 342–43n, 353n
in Earth's early atmosphere, 24–25, 42
on exoplanets, 312
light emitted by, 311
liquefaction of, 348–49n
molecular atmospheric dispersal of, 196, *196*
and nuclear weapons, 256
hemoglobin, 116, 341–42n
Hero of Alexandria, *339*, 340n
Herschel, William, 170, 305, 306
Hiroshima, Japan, 42, 235, 238, 242, 247, 249, 299, 326, 348n
Hitler, Adolf, 71–74, 109
Hooke, Robert, 115, 269
horsepower standard, 173–74
human flight, 13, 197–203, 205–08, 222
hurricanes, 274, 276–84, 291–92
hydrochloric acid (HCl), 66
hydrogen (H₂)
and ammonia production, 51, 57–58
in balloons, 199–200, 202, 205–06
and coal gas, 224
discovery of, 85, 89
on exoplanets, 312
and flatulence, 155
and formation of solar system, 23, 321
and ideal gas law, 209–10
and lighting, 224–27, 311
and nuclear weapons, 256
as percentage of life forms' weight, 50
polarity of, 138
reactivity with argon, 216
and refrigeration, 260
and space colonization, 323

Hydrogen sulfide (SO₂), 17, *17*, 25, 30, 156, 312, 324
hypochlorous acid (HClO), 66

ice crystals, and rain-making, 272–73, *272*, 275, 275–78
Industrial Revolution, 13, 159, 163, 172, 175, 184, 223, 315
infrared light, 314–15
iodine-131 (I), 231, *231*, 254–55
iron
and Earth's formation, 24
and evolution of Earth's atmosphere, 28, 99–100
in hemoglobin, 116
oxidation of, 79, 335n
and solar system formation, 23
and steel production, 187–90, 194
Tay Bridge's low-quality, 192

Jackson, Charles Thomas, *338*, 338n
Janssen, Jules, 342–43n
Jenner, Edward, 129–31
Johnston, David, 32, 37
Jupiter, 23, 308

Krakatoa, 42, 319, 351n
krypton, 217–18, 227

Lake Nyos, Cameroon, 42–48, *45*
Langmuir, Irving, 268–80, *272*, 282, 284
latent heat, 168, 262–63, 271, 278
laughing gas. *See* nitrous oxide (N₂O)
lava, and early Earth's atmosphere, 24–25
Lavoisier, Anne-Marie, 91–93, 95–96
Lavoisier, Antoine-Laurent
burning diamonds with sunlight, *92*, 93
and conservation of mass law, 97, 99, 104
death of, *91*, 109–10
and discovery of oxygen, 82, 91, 94–96, 104, 110
and Ferme Générale, 92, 97, 108–09
and French Revolution, 107–08
and gunpowder, 336n
and melting of metals, 77, 79

Lavoisier, Antoine-Laurent *(cont.)*
  study of gases by, 93, *95*, 97–98, 198,
    200, 209, 212
law of gases, 206–10, 292, 350n
Lewes, George, 112–14, 116–17
life
  and climate change, 317
  emergence of, 13, 304
  and evolution of Earth's atmosphere, 15
  extraterrestrial, 304–09, 312–14,
    353–54n
  and nitrogen production, 50, 54–55,
    99, 331n
  and oxygen production, 15, 98–104,
    355n
  and ozone, 297, 303
light bulbs, 58, 226–27
lighting, 13, 223–28, 310–12
lightning, 54, 56–57, 350n
liquids, compared to solids and gases,
    23–24, 71, 168, 262
Long, Crawford, 143–44
Lorenz, Edward, 287–92
Los Alamos, New Mexico, 236, 239–42,
    345n
Louis-Philippe (king of France), 146
Louis XVI (king of France), 201
Lunar Society, 105, 124, 170, 224

MacArthur, Douglas, 238–39
Macbeth (Shakespeare), 186, 341n
magma, 25–26, 28–31, 34
magnesium, 132, 214–15
Manhattan Project, 61, 231, 236, 238–44,
    243, 249, 302, 333–35n, 345n
manure, 55–56
Mars, 305–07, 315, 320–21, 340n, 355n
Maxwell, James Clerk, 221, 326
McGonagall, William Topaz, 185–87,
    186, 194, 341n
Mendeleev, Dmitri, 216, 218–19, 343n
mercury, 25, 213–14
mercury cyanide, 87
metal, cutting and welding, 75–80, 94,
    335n
meteorology
  and chaos theory, 222, 291–92

James Espy's experiments, 270, 274,
    281
  and ideal gas law, 350n
Irving Langmuir's weather control
    experiments, 268–72, *272*, 274–80,
    282, 284
  and nuclear fallout, 257–58
Project Cirrus, 276–80
Project Popeye, 281–84
weather forecasting, 284–91, *285*,
    293–96, *294*
methane (CH_4)
  and coal gas, 224
  and cutting and welding metal, 76–79
  on exoplanets, 312
  and extraterrestrial life, 309
  and flatulence, 155–56
  heat absorption by, 315–17, 354–55n
  and lighting, 223–24, 227
  molecular atmospheric dispersal of,
    75, *75*
  and oxygen, 101
  and refrigeration, 260, 263–65
methanethiol (CH_3SH), 151, *151*, 156
methanol, 265
methyl chloride, 262–63
mineral formation, and oxygen, 101–02
mitochondria, 101
Montgolfier, Jacques-Étienne, 198–99,
    201, 206, 211, 342n
Montgolfier, Joseph-Michel, 197–99,
    201, 206, 211, 342n
moon (Luna), 26–28, 100, 305–06,
    320–21, 355n
Morton, William, 133, 135–44, *136*, 338n
Mount Saint Helens, Washington
  eruptions of, 18, 21–22, 29, 31, 35–42,
    36, *38*, 299
  pre-eruption evacuations of, 31–35
  Harry R. Truman's experiences, 18–22,
    19, 29, 32–35, 38–41, 326, 329n
Mount Vesuvius, 38–40
Murdoch, William, 174, 225
mustard gas, 67, 333n

Nagasaki, Japan, 233, 235, 238–39, 242,
    247, 249

neon, 218, 227–28, 311
Nernst, Walther, 52–53, 56, 58, 64, 72
neutrons, in nuclear reactions, 237,
    240–41, 243, 250, 267
Newcomen, Thomas, 163, 166–70, 172,
    266
nitric oxide, 183, 260
nitrogen (N_2)
    and ammonia production, 50–54,
        56–62
    and bacteria, 54–55, 99, 331n
    compared to water vapor, 270
    discovery of, 85
    and evolution of Earth's atmosphere,
        48, 53–54, 82, 98, 102, 119, 331n
    heat absorption by, 314–15
    and ideal gas law, 210
    in lightbulbs, 227
    molecular atmospheric dispersal of,
        49, 49, 207
    and nuclear reactions, 241
    and plate tectonics, 30
    polarity of, 138
    and refrigeration, 260
    and space colonization, 323
    John Strutt's study of, 212–15
    triple bonds of, 53–54
    volcanoes' production of, 50, 53,
        331n
nitrogenase, 54–55, 100
nitrogen oxides, 250, 318
nitroglycerin, 115, 175–81, 183, 341n
nitrous oxide (N_2O)
    as anesthesia, 133–38, 142, 146
    Thomas Beddoes's experiments with,
        122, 123, 127–28, 131–32, 134
    Humphry Davy's experiments with,
        127–29, 131–33, 326
    discovery of, 89
    heat absorption by, 317
    molecular atmospheric dispersal of,
        121, 121
    and space colonization, 324
Nobel, Alfred, 177–84, 179, 219, 238,
    256
Nobel, Immanuel, 177–78
Nobel, Oscar-Emil, 178–79

Nobel Prizes, 182–84, 219
noble gases, 213, 218–19, 223, 227, 252,
    343n
nuclear chain reactions, 236, 239–41,
    243, 267
nuclear weapons
    after-effects of, 238–39, 247
    and atomic physics, 222
    Baker test, 246–47, 246, 255–56
    Castle Bravo test, 256–57, 347–48n
    fallout from, 235–36, 238, 248–51,
        248, 250–55, 257, 284, 296, 300,
        346, 346–47n
    Gilda/Able test, 234, 246–47, 255
    government propaganda for, 235
    Operation Crossroads test, 231–36,
        232, 242, 245, 254
    protests against testing of, 251
    public apathy towards, 247–48, 250
    and refrigeration, 259
    and terraforming, 320
    testing of Supers, 256–57
    testing survivability after detonation,
        248–49, 346, 346–47n
    Trinity test, 237–39
    in USSR, 296, 300, 302–03

obstetrics, 145–46
oceanic sound channels, 351–52n
oganesson, 342n
Operation Crossroads nuclear test,
    231–36, 232, 242, 245, 254
Oppenheimer, Robert, 238, 345n,
    345–46n
osmium, 58, 60
oxidation, 79, 94, 335n
oxygen (O_2)
    and aerobic bacteria, 100–101
    and ammonia production, 57
    and Arizona biosphere experiment,
        355n
    and burning metal, 75–80, 94, 335n
    and carbon-14, 253
    compared to water vapor, 270
    and CO poisoning, 127, 341–42n
    created by living things, 15, 98–10
        355n

oxygen (O_2) (cont.)
  discovery of, 81–82, 86–87, 87, 87, 89–91, 91, 94–96, 104–10
  and evolution of Earth's atmosphere, 75, 82–83, 99–103, 119
  on exoplanets, 312–13
  heat absorption by, 314–15
  in human metabolism, 112, 123–24
  and lighting, 225–27
  molecular atmospheric dispersal of, 81, 81, 207
  as percentage of life forms' weight, 50
  from photosynthesis, 98
  and plants, 101, 336n
  polarity of, 138
  and purification of nitrogen, 212
  reactivity with argon, 216
  and refrigeration, 260–61
  and respiration, 94–96, 103–04, 116
  and solar system formation, 23
  and space colonization, 323–24
  and steel production, 190, 194
Oxygen Catastrophe, 13, 82, 188
ozone (O_3), 293, 293, 297–99, 303, 312, 317, 324

Pell, Claiborne, 282–83
periodic table of elements, 12, 24–25, 60, 82, 216–19, 251, 311, 343n
permanent gases, 260, 348–49n
Petty, William (Lord Shelburne), 89–90
phlogiston theory, 89–90, 93–96, 104
phosgene gas, 67, 333n
phosphorus, 191, 216
photosynthesis, 98–99
phytoplankton, 319
Pictet, Raoul, 348–49n
Pilâtre de Rozier, Jean-François, 201, 203
Planck, Max, 72–74
plasma, 237, 306
plate tectonics, 29–31
platinum, 322
plutonium, 236–37, 239–40, 243, 245, 247, 249–50, 256, 345n
Pneumatic Institution, 124–33
polarity, and volatility of liquids, 138, 145
pollution, 228, 313–14, 317

polonium, 237
pop culture, and nuclear testing, 251, 347n
potassium, 132, 216
potassium-40, 252, 312
pressure
  and ammonia production, 57–58, 60–61
  of Martian atmosphere, 307
  and nitroglycerin explosions, 178
  and pumping water, 166
  relationship with temperature/volume in gases, 24, 26–27, 31, 152, 206–10, 292, 322
  and transformation of gases, 24
Priestley, Joseph
  and discovery of oxygen, 81–82, 87, 87, 89–91, 91, 94–96, 104
  and French Revolution, 105, 107–10
  and Lunar Society, 124, 170, 224
  Priestly Riot, 105–06, 106
  study of gases by, 88–90, 107, 198, 209, 222, 335–36n
  as subject of caricature, 88, 131
  and James Watt, 170
Project Cirrus, 276–80
Project Mogul, 300–303, 301
Project Popeye, 281–84
Project Stormfury, 280, 283–84, 290
proteins, 54, 62, 99
Pujol, Joseph "Le Pétomane," 152–57, 154

quantum mechanics, 222

radioactivity. See also nuclear weapons
  and Baker nuclear test, 246–47
  and cancer, 249, 253–55
  deadly accidents with, 239–45, 243, 256
  disregard for dangers of, 238–39, 247–49, 248, 345–46n
  and D. radiodurans bacterium, 309
  in fallout, 250–55, 257, 284
  natural sources of, 252–53
  and Operation Crossroads, 235–36, 254
  and plants, 251–54
  of plutonium, 236, 240, 243, 247
radon, 210, 218, 252
rain-making, 268–83, 272

Ramsay, William, 213–19, *213*, 222, 343n
Ray, Dixy Lee, 33, 35
refraction of sound, 297
refrigeration
  absorption refrigeration, 263–65
  and chlorofluorocarbons, 266,
    349–50n
  Albert Einstein's work on, 12,
    259–60, 263–66, *264*
  and gas compression/liquefaction,
    260–66
  and nitrogen/ammonia production, 60
  and nuclear weapons, 259
religion, and extraterrestrial life, 353–54n
respiration
  and atmospheric dispersal of gas
    molecules, 3–4, 7–11, 17, 43, 49, 75,
    111, 121, 151, 158, 185, 196, 223,
    231, 259, 293
  and blood circulation, 115–16
  as burning, 82, 90, 96, 115–16
  Caesar's last, 7–11, 41, 326, 351n
  and CO_2, 84–85
  and gas warfare, 66, 68
  Antoine-Laurent Lavoisier's studies
    of, 94–95, *95*
  and natural radioactivity, 252
  and nuclear fallout, 195–36
  and oxygen, 94–96, 103–04, 116
Richardson, Lewis Fry, 285–88, *285*, 292
Robert, Anne-Jean, 199, 201–02
Robert, Nicolas-Louis, 199, 201
Roswell, New Mexico, 14, 293–96, *298*,
  303, 305
rust, 79, 100, 335n

Saturn, 23, 308
Savery, Thomas, 164–66, 264
Schaefer, Vincent, 271–73, *272*
Scheele, Carl, 86–87, 90, 96
seeding clouds, 268, 270, 273–75, 277,
  279–84, 292
Shakespeare, William, 186, 341n
silicon, 23, 28
silver iodide (AgI), 268, *268*, 275, 277,
  279–83
Simpson, James, 144–45

sky, blue color of, 203, 219–22, 319, 321,
  326, 344n
Slotin, Louis, 237, 242–45, 249, 256
Sobrero, Ascanio, 175–76
sodium
  discovery of, 132, 337n
  and lighting, 227, 311
  and refrigeration, 265
sodium chloride, 28
solar system, formation of, 22–23, 27, 321
solar wind, 24–25, 307
solids, compared to liquids and gases,
  23–24, 71, 168, 262
sound. *See also* acoustics
  of volcanic eruptions, 35–37, 41
sound channels, 296–300, 351n,
  351–52n
space colonization, 13, 323–26
Spirit Lake, Washington, 19–21, 38
spontaneous human combustion, 13,
  111–12, *113*, 116–17
steam
  in balloons, 198–99, 205
  driving machinery, 163–64, 166–75,
    184, 223–24, *339*, 340n
  in Earth's early atmosphere, 26
  energy required to produce, 168–69
  Thomas Newcomen's engine, 163,
    166–70, 172, 266
  and vaporization, 39–40
  in volcanic eruptions, 31
  James Watt's engines, 124, 163, 167–75,
    171, 191, 225, 262, 265, 340n
steel, 184, 187–92, 194–95
stratosphere, 208, 266, 293
strontium, 132, 231, *231*, 251, 254–55
Strutt, John William (Lord Rayleigh),
  211–17, *213*, 219–21, 269, 344n
sulfur
  and anaerobic bacteria, 98
  as chemical weapon, 64
  and coal gas, 226
  and gunpowder, 175
  in hell, 352–53n
  in nitrogenase, 54
  reactivity with argon, 216
  from volcanoes, 37, 46, 98

Sulfur dioxide (SO$_2$), 17, *17*, 25, 30, 34, 89, 262, 318–19, 331
sulfuric acid, 200, 212
sun, 23, 256, 306, 311, 321–22, 326, 353n
sunlight, 98–99, 220–21, 314, 319, 321, 354n
supercooled water, 271, 272–73, 275, 277, 283
surgery, 133–34, 139–42, *141*, 145
synthetic gasoline, 71–72
Szilard, Leo, 260, 263–67, *264*

Tambora, 42, 319
Tay Bridge, 192–94, *193*
temperature
    and acoustics, 296–300
    and ammonia/nitrogen production, 52, 57–58, 61
    and emission/absorption spectrums, 311
    and greenhouse gases, 315, 354n
    and heaven and hell, 352–53n
    and Lake Nyos's CO$_2$ eruption, 46
    and lighting, 227–28
    and liquefaction of gases, 260–62
    and plate tectonics, 30
    and rain-making, 271, 277
    relationship with pressure/volume in gases, 24, 26–27, 31, 152, 206–10, 292, 322
    and state of matter, 97, 168
    of the sun, 321–22
    and transformation of gases, 24, 26–27
terraforming, 320–21, 355n
Theia, 27–28
theory of relativity, 266–67, 343n
Thoreau, Henry David, 338n
Trinity nuclear test, 237–39
Truman, Edna "Eddie," 20–21, 29, 32
Truman, Harry Randall, 18–22, *19*, 29, 32–35, 38–41, 326
tungsten-carbide, 240, 243
turbulence, 287, 292, 351n

ultraviolet light, 297–98, 303, 312
unidentified flying objects, 295, *298*, 302
uranium, 217, 236, 267, 345n

vacuum
    and air pressure, 159–65, 167
    and buoyancy of fluids, 205
    and flatulence, 153
    and lightbulbs, 227
    pumping water with, 164–67, 340n
    and refrigeration, 265
    of space, 323
    and subatomic particles, 339n
    Otto von Guericke's studies of, 158–63, *162*, 172
van Helmont, Jan Baptista, 83, 105
vaporization, 39–41, 234, 237, 246
Venus, 305, 307–08, 315, 317, 340n, 352n, 354n
Vietnam war, 281–83
volatility of liquids, 138, 140, 145
volcanoes. *See also specific volcanoes*
    and emergence of life, 98
    on exoplanets, 312
    and formation of Earth's atmosphere, 13, 15, 24–26, 28, 31, 42, 48, 50, 53–54, 307
    Icelandic eruptions, 329–31n
    and Lake Nyos's CO$_2$ eruption, 46–47
    and Mars's atmosphere, 307
    and nitrogen production, 50, 53, 331n
    and sulfur dioxide, 319
    sulfur from, 37, 46, 98
    vaporizing heat blast from, 39–41
    and Venus's atmosphere, 307
volume, relationship with pressure/temperature in gases, 24, 26–27, 31, 152, 206–10, 292, 322
Vonnegut, Bernard, *272*, 275–76

Warren, John, 139–42
water (H$_2$O)
    on asteroids, 322–23
    buoyant force of, 204
    and CO$_2$, 26
    driving machines with, 164, 340n
    and extraterrestrial life, 309
    formed from flammable gases, 86, 96
    and greenhouse gases, 315
    molecular atmospheric dispersal of, 158, *158*

O–H bonds of, 329n
polarity of, 138
pumping, 164–67, 340n
and rust, 335n
supercooled, 271, 272–73, 275, 277, 283
water pressure refrigeration, 265
water vapor, 25, 39–40, 270–71, 274,
    292, 307, 311–12, 315, 317
Watt, James
and horsepower standard, 173–74
and Lunar Society, 105, 124, 170
and meteorology, 269
and patent rights, 174, 181
and Pneumatic Institution, 124–25, 129
and steam engines, 124, 163, 167–75,
    171, 191, 225, 262, 265, 340n

Watt, Jessie, 124, 170
weather. *See* meteorology
Wegener, Alfred, 29–30
weight, of gases, 13
welding, 75–80, 94, 335n
Wells, Horace, 133–37, *135*, 144, 146–47
World War I, 11, 63–65, *65*, 72, 157, 219,
    250, 344n
World War II, 56, 72, 74, 271
wrought iron, 187–88, 190, 192. *See
    also* iron

xenon, 218, 227

Yellowstone eruptions, 42, 329–31n
Ypres, France, 65, *65*, 69

About the Author

SAM KEAN is the *New York Times* bestselling author of *The Tale of the Dueling Neurosurgeons, The Disappearing Spoon,* and *The Violinist's Thumb,* all of which were also named Amazon top science books of the year. *The Disappearing Spoon* was a runner-up for the Royal Society of London's book of the year for 2010, and *The Violinist's Thumb* and *The Tale of the Dueling Neurosurgeons* were nominated for the PEN/E. O. Wilson Literary Science Writing Award in 2013 and 2015, as well as the AAAS/Subaru SB&F prize. Kean's work has appeared in *The Best American Nature and Science Writing, The New Yorker, The Atlantic,* the *New York Times Magazine, Psychology Today, Slate, Mental Floss,* and other publications, and he has been featured on NPR's *Radiolab, All Things Considered,* and *Fresh Air.*